VISITORS FROM ANOTHER DIMENSION

(A Revised and Expanded Edition)

By
James L. Eakins
Ozark, Missouri

Visitors From Another Dimension
Copyright © 2024 by James L. Eakins
ISBN 979-8-9920492-3-7
Second Printing 2025
Revised and expanded edition.

A Novella - a realistic fiction story

All rights reserved.
Reproduction of text in whole or in part
without the express written consent of the author
is not permitted and is unlawful
according to the 1976 United States Copyright Act.

Published by:
James L. Eakins
Ozark, Mo 65721

Printed in the United States of America

Irwin Printing Company
260 NE US Highway 60
Billings, Missouri 65610

DEDICATION

To my creator, the Lord Jesus Christ. Nothing was made without Him. He was in the world, and the world was made by Him. For by Him all things were created, in heaven and on earth, visible and invisible, whether they are thrones, dominions, principalities, or powers: all things were created by Him and for Him, and He is before all things, and by Him all things consist.

J.E.

ACKNOWLEDGMENTS

Above all, I want to express my deepest gratitude to my Lord and my God, Jesus Christ, who constantly motivates me with the Holy Spirit's prompting.

Forever filled with gratitude,

-James

ABOUT THE AUTHOR

James L. Eakins was born on December 4, 1953, to Guy and Martha Jane Eakins in the small town of Ozark, Missouri, where he spent his childhood years and attended school. In 1975, he married Judy (Johnson) from Galena, Missouri, and they eventually became the proud parents of seven beautiful children: Galen, Misty, Rachel, Benjamin, Joshua, Charity, and Judy Kaylee.

Jesus Christ transformed his life in 1978 and was immediately called to lead souls to His Lord. He began a personal ministry of soul-winning, going door-to-door, preaching the gospel, and distributing gospel tracts. Responding to God's call to preach, he held his first revival later that year, marking the start of his evangelistic travels within a 250-mile radius of his home, preaching in any church where God opened the door.

James began his pastoral journey at McCord Bend Southern Baptist Mission Church. He was later led to Wheelerville Union Church, which grew until it overflowed. An old-fashioned Brush Arbor was constructed at Scholten Corner, not far away, to accommodate the expanding congregation. Night after night, the Spirit of God moved powerfully, igniting a revival that swept through the countryside and transformed countless lives for eternity. Eventually, God guided him to purchase a storefront building on Main Street in Crane, Missouri, where he founded New Life Fellowship Church in 1985.

As cassette tapes gained popularity, God provided

him with a means to record and duplicate his sermons for widespread distribution. He began offering these free sermon tapes at his church, during his revivals, door-to-door, and by mail upon request. This outreach generated numerous phone calls inviting him to preach in various locations. In 1987, Touching the World Ministries was established, marking the beginning of a new era of travel and adventure. He preached in different churches across the United States, including many Native American Indian Reservation churches. His ministry also took him to Central America, where he led a significant crusade in Belize City and held revivals in Dangriga and the surrounding areas, transforming the lives of countless individuals. Feeling called by God to provide more consistent and in-depth ministry to the Native American community, rather than making occasional visits, he relocated his family to Pyramid Lake, in Nixon, Nevada, in 1990, where he ministered and served the Indians on the Paiute reservation. He later spent three years pastoring the Calvary Assembly of God Church in Gardnerville, Nevada, situated in the Carson Valley. During this time, he embarked on a new endeavor of proclaiming the gospel through the printed word. The church produced a multi-page newspaper, published it, and distributed it free of charge.

Returning to his hometown in 1994, he founded Ozark Full Gospel Church, where he has served as pastor to this day. Since its establishment, he has initiated numerous programs, including a Christian school, five revival centers, several radio broadcasts, and the Ozarks' Good News newspaper. His sermons air on television every Sunday morning and are also available on CDs, DVDs, podcasts, and live streams, as well as on Facebook and YouTube. He is an avid reader and Bible student, making him a well-known self-taught, self-inflicted Bible scholar recognized for his in-depth teaching and preaching; hundreds of thousands of his messages continue to be distributed worldwide. James is the author of numerous articles, sermon booklets, and the books Run Chicken Run, Brush with Glory, Unique Stories from My Ozark Childhood, My Name Is Adam, My Name Is Methuselah, My Name Is Noah, and Simple Poems of Life.

Don't proofread my life
unless it is for a
glorious publication.
~ James L. Eakins

PREFACE

Considering the growing presence of artificial intelligence in our world and the impending emergence of superintelligence, along with the rapid advancement of human technology, it is unsurprising that visitors arrive from another world, bringing with them a powerful delusion that we might accept as truth. They may or may not come by spacecraft; they might approach us through time travel in dark space, or they could originate from another dimension.

This book is fiction, but a story similar to its plot could actually happen someday. We must never forget that God is real and in complete control, so hold tight to your faith and to who God is. The God of the Bible cannot be disproved. From everlasting to everlasting, Jesus is Lord.

J.E.

INTRODUCTION

This book is both alarming and straightforward. It discusses the real possibility of a strong delusion that could lead all mankind to believe a lie. Given the prevalence of superintelligence and super-gullibility in our world, we may indeed be visited by beings from another planet. We are both warned and encouraged to maintain our faith in God in a world that, for the most part, doesn't honestly believe in a creator. This book may become a manual for how true believers should respond to visitors from another world. You will see that nothing disproves the God of the Bible or the God revealed in His word. This revised and expanded edition tells a compelling story of how a preacher confronts one of the most significant challenges in his ministry.

-A faithful friend of God.

VISITORS FROM ANOTHER DIMENSION

(A Revised and Expanded Edition)

James L. Eakins

IT FELT LIKE ANY OTHER DAY

It was like any other day; winter had arrived, and the temperature was freezing. But like every morning, I started my car to warm it up. Then, I had breakfast and quickly proceeded outside to my warm car. I backed out of my driveway to head to my church office. Out of habit, I turned on the car radio, and to my shock, it was blaring. "Now, we know we are not alone; we have visitors from another distant planet." I thought this was a fictional news report, so I turned the radio dial again and again; every station was giving a special news report. Instantly, I thought this was a big hoax; my heart and mind were flooded with disbelief and confusion. Surely, this is just a big lie; it just can't be true.

As I finally arrived at the church, I felt numb from the news of visitors from another planet. In a daze, I pulled into the church parking lot, which was full of cars, all with their radios blaring. Many people stood fearfully outside their vehicles, while others crowded at the church doors. Inside the church, it was packed. Not knowing most of these people, I hurried inside, squeezing through the crowd. After all, this is my church, the place where I serve as pastor.

As I approached the altars at the front of the sanctuary, I saw my Christian brothers and sisters. They were praying, and it was clear that the report of visitors from another world had shaken them. Then my gaze settled on my good friends Terry, Don, Chuck, and Doc. It was evident they were seeking answers, as reflected in the confused expressions on their troubled faces. Chuck then asked me the million-dollar question, "Preacher, are these beings demons masquerading in bodies to disrupt the gospel of Christ?" I answered, "No, Chuck, these are not demons; they do not have bodies. These may be fallen angels who can assume a physical form. But let's not jump to conclusions yet; let us wait and see."

Regardless of who these visitors may be, my faith in God has never been stronger. Regardless of who they are or what their intentions might be, Jesus Christ is Lord, and the Bible is the absolute truth. If these visitors are indeed people from another planet, our God is still their creator. If they don't know who Christ is, it would be wise to tell them. However, I am pretty sure they already know the truth. It's possible that they are just a distraction or a deceiving lie that's coming. Whatever this is, it only makes our God bigger and greater than we have ever imagined. My faith in God is greater now than ever before.

Due to the overwhelming crowds in and around the sanctuary, Josh and his team of worship and praise leaders spent many hours engaging in music and worship. I then preached for an extended period as the people craved more, and the crowds continued to grow. Given the increasing number of attendees, I decided to hold services continuously for sixteen hours a day, every day, without pause. We were open for service from 6:00 a.m. to 10:00 p.m.

We rotated our music teams and preachers. Pastor Josh, Jimmy, Chris, Tyler, Logan, and I preached from the Bible every day. News of these visitors spread quickly, and the crowds continued to come for comfort and answers. We encouraged the masses to trust in our one true God.

And for this reason,
God will send them
strong delusion
so that they will believe a lie:
that they all might be
condemned who did not
believe the truth
but took pleasure
in unrighteousness.

GOING TO INVESTIGATE

It was time for me to go and investigate these visitors from another world. I heard that these visitors wanted to hold town hall meetings with the people because they didn't trust the news networks or social media. They insisted on meeting the people in person and also didn't trust our artificial intelligence or any of our technology. They continued to insist on face-to-face meetings in theaters, concert halls, school gymnasiums, sports stadiums, and even churches. They were scheduled to have a town hall meeting in Chicago, Illinois, so I decided to meet them there.

There weren't any planes available because all flights had been suspended. It's strange that the visitor's spacecraft from this so-called other planet was nowhere to be found. While driving to Chicago, I called Pastor Josh to check in and see how things were going at the church. He mentioned that the crowds hadn't let up, and as in every service, many were being saved from their sins and bondages by the power of the gospel of Christ. I encouraged Josh not to give up and urged him to inform the people that I would find some answers and report back to them as soon as possible. It was my duty to seek answers, while it was Josh's responsibility to maintain the health of our church.

The traffic in Chicago was a nightmare, but my car

was reliable, and I finally arrived at the United Center ten hours ahead of schedule. I wanted to ensure that I was inside the meeting place to hide, hear, and observe these strange, amazing beings. After hours of waiting, I finally got inside and took a seat. It was then that I met Doctor Lucas, Loucuss, a pastor of a megachurch in Atlanta, Georgia. Lucas told me that his first and last names are spelled differently but pronounced the same, so feel free to call me Lucas, and I'll sort it all out in my head. During my wait, I also met Pastor Bobby Cain and Pastor Tommy Demas, both from Dallas, Texas. Pastor Demas told me he had heard on the news that these visitors were scattered all over the world, primarily in the United States. Without exception, they insisted they didn't trust any social media but wanted to meet people in person.

Just minutes before the meeting was about to commence, Pastor Josh called. He had a good report and said the crowds there were not letting up; many people were coming to a proper knowledge of Christ.

Before hanging up, Josh informed me that these strange beings were coming to Springfield, Missouri, in about a month. I promptly told him to instruct the people at our church not to meet these beings, as we do not know enough about them, and that their time would be better spent hearing solid Bible truth.

Josh also told me that many pastors and churches claimed these visitors were angels sent by God to save the world. **Some pastors even prophesied and preached that these were holy visitors—angels sent**

from God, a manifestation of God's movement to further the gospel and prepare the way for a great revival. Of course, Pastor Josh quickly debunked this lie, telling the crowds that when God's Holy Angels come, it will be to execute the wrath of God, not to save the planet, but to judge the ungodly. In times past, angels have helped individuals along the way, including our Lord. However, God's word is clear that busy angel activity at the close of this age will not be a welcoming sight. These visitors before the flood of Noah were evil, and they will be evil before the second coming of Jesus Christ to this Earth.

My last instructions for Pastor Josh were to warn the congregation not to listen to or watch any news or social media, as it was impossible to discern what was real and what was not.

Jesus Christ is
neither recognized
nor respected
in the plans
of these visitors
from another world.

THE MEETING HAS BEGUN

They didn't start the meeting with music but rather with praise for the people of Earth, telling us that we were a great and mighty people who deserve the very best from their world and ours. Their appearance was stunning and radiant. They were very handsome men, about 8 feet tall, towering over us physically, but only in that regard. They were witty and charming, without any hint of arrogance. They told us they wanted to meet each of us in person to win our hearts and earn our trust during town hall meetings held in school gymnasiums, stadiums, churches, and similar venues. They mentioned they had no need to eat, but if we insisted, they would join us for a meal and fellowship, if that is one of your essential customs.

These visitors from another world warned us that our artificial intelligence cannot be trusted, at least not on its current trajectory toward superintelligence. All forms of media and communication must be reprogrammed to function cohesively without divisions or schisms. They stated that, for our safety, they would take complete control of all our communications, internet, and computer functions. However, we could still maintain limited cell phone communication for banking, buying gas, food, and other essential items, which allows us to stay in touch with friends and family. Our debit and credit cards will continue to function as usual. We can still access the internet through our computers and devices, enabling us to upload and download, but under constant surveillance. Hate speech and religious rhetoric will not be tolerated in any form. However, they reminded us that there will always be the good, the bad, and the ugly in cyberspace.

They cheerfully declared that we have miracles concerning your harsh weather, earthquakes, tsunamis, diseases, plagues, premature death, and aging. We can prevent hurricanes, tornadoes, wildfires, and other natural disasters. (They kept saying, "Believe it or not, we're not lying.") It was clear they were not from Earth, yet they never disclosed their origin. Someone asked them, "Where is your security?" They replied that there is nothing on Earth that can hurt us.

The first reports of these beings from another planet emerged in our news and special reports from the Pentagon. The big news was that we are not alone; there is life beyond our planet.

The fact that these beings constantly reiterated that we are not lying raised a red flag for me.

There was no pomp and circumstance, and none of our politicians were acknowledged; it felt as if they were muted and deliberately avoided. A loud heckler shouted from the back, and one of the strange beings, while speaking, raised his hand to signal for silence, causing the entire stadium to fall quiet; no one could hear anything except the peculiar visitor's voice. These remarkable beings were in complete control.

ARE THEY TIME TRAVELERS?

In my spirit, I felt they couldn't be a highly advanced race from another planet. Maybe they are time jumpers or time travelers, or my worst fear, fallen spirit beings taking on human form. If they are visitors from another planet, it does not shake my faith in God. If they are the return of fallen angels, it powerfully stirs my faith in God. If they are indeed time travelers or fallen spirit beings, then they are up to evil. If they are truly visitors from another planet, we need answers, and so do they. As you're reading my thoughts, I'm sure you can tell I'm still somewhat confused. But my faith in God has never been stronger.

It's now late in the day, and it's been a long, exhausting day. Finally, the meeting is over. I haven't had anything to eat or drink, and I am starving. As I'm leaving, one of these strange beings came towering over me and introduced himself as Lucian. Somehow, he already knew I was a preacher, though I hadn't told him. He asked if I enjoyed the presentation and if I had any questions. I replied, "I do have a question. Do you believe in God?" He answered, "I believe and I personally know God doesn't, everybody?" Then I asked the million-dollar question fearfully. "Do you personally know Jesus Christ?" He never answered me; he just hurried away.

Shortly, I gathered with Pastors Demas, Lucas, and Cain. They were excited, to say the least; it was clear they believed every word from these beings. Lucas, the megachurch pastor from Atlanta, Georgia, mentioned that he was eager for these amazing beings to visit his church. I didn't say a word. I then invited them to dine with me at the nearest restaurant. When we were seated at the table, the atmosphere around us buzzed with chatter about these visitors from another planet. Pastor Cain exclaimed excitedly that even the President of the United States endorses them. But I said, 'Let's be cautious and look at what the Bible says or doesn't say.' The President of the United States is not our commander-in-chief, Shepherd.

Pastor Cain was a vegetarian, so he ordered a salad and steamed broccoli. Pastor Lucas ordered a ribeye steak with all the fixings and a stuffed baked potato. Pastor Demas chose pork chops, mashed potatoes, and gravy. By this point, I had lost my appetite, so I just ordered some iced tea to drink. This day was weighing down on me like a ton of bricks.

Then I informed my new pastor friends that I do not believe these visitors are highly advanced travelers from another planet. If they are time travelers, they are up to no good. Pastor Demas, Lucas, and Cain shouted across the table at me. You're just overreacting; I wondered where this behavior came from. I then proceeded to tell them that these visitors, wherever they have come from, are highly advanced liars and part of an end-time deception. At that moment, I no longer had three new friends.

As we departed the restaurant, I reminded my new-found friends whom I had just now lost. That today's meeting was a rally meeting, and not once was there a prayer or a mention of God. Remember, they said they did not travel by brute force, but by a highly advanced, powerful, and silent light.

Lucas asked if I would at least come to his church in Atlanta, Georgia, when these strange, charming visitors are scheduled to be there with him. Lucas said I could investigate a little more and perhaps gain a better perspective. Reluctantly, I agreed and we exchanged phone numbers.

This doesn't surprise us
because even Satan
can disguise himself
as an angel of light.
So, it's no surprise
that his servants
also try to pose as
ministers of righteousness.
But in the end,
they will be exposed
and receive
what they deserve.

DRIVING IT HOME

Very early the next morning, I began driving home, while God drove everything home into my heart and spirit. This is the great acceleration of a strong delusion to believe a lie.

As I reached for my phone, it struck me that all I had to do was tell my car dashboard to call Pastor Josh, and it would respond with, "Calling Pastor Josh." This technology is impressive; my car can even drive itself and parallel park without my help. But superintelligence will soon spiral completely out of control. These visitors from another world have remarkable plans for our Artificial Superintelligence. Josh answered, and I told him I had answers for our church. I spoke with Josh for about an hour, but he was interrupted by others. I told him that when I returned, he should prepare for some lengthy sessions from the pulpit. He informed me that the crowds were gathering nonstop, and the gospel of Christ was still inspiring people.

Arriving home, it was around seven p.m. I immediately went to the church. The church parking lot was filled with cars, and inside, the church was packed from the fellowship hall through the sanctuary. The people were singing "All is Well with My Soul" and "God How Great Thou Art."

After the worship service, I asked everyone to please take their seats. Then, I began by thanking Pastor Josh and his preaching team for keeping the pulpit of God sanctified and as a holy place for God's Word. I also expressed gratitude to all the dedicated musicians and worship leaders. I continued by thanking everyone for attending and shared that we are truly not alone; God is our refuge. We must hold tight to one another in the truth of God's written Word and never forsake gathering together. We must not rely on newly printed materials, media, or information obtained through technology, especially artificial intelligence. We need to meet face-to-face, hand in hand, and engage in heart-to-heart conversations centered on God's written Word. Gathering around the Bible is our only chance for our families to stay safe. We can no longer clearly distinguish what is real from what is not. Our tangible Bible is what we can trust until the coming of our Lord Jesus Christ. After preaching for two hours, we all prayed for over an hour and then closed the night by singing the old hymn "Hold to God's unchanging hand."

Hebrews 10:25 KJV states:

"Not forsaking the assembling
of ourselves together,
as the manner of some is;
but exhorting one another:
and so much the more,
as ye see the day approaching."

The Book

Our eyes cannot focus
in the thick darkness of night.
Nor can we bear to stare
into the burning sunlight.

Do we close the Book of Life
and pretend?
Or do we leave the book open
and be faithful to the end?

As we travel through
the day and night.
The book remains open
because the Author still writes.

~ James L. Eakins

THE END-TIME

The next morning, I arrived at the church, which was on Thursday. The Church sanctuary was already full, and the praise and worship leaders were singing in a lovely spiritual flow. I asked Pastor Josh if he would preach about the creation, the preexisting Christ, and the Word made flesh. Which he did, and it was so good. He preached for about an hour, and the crowds loved it; they couldn't get enough of God's word.

It was now my turn to preach the coming of Jesus Christ. But first, I wanted to expose these so-called visitors from another planet. They are not highly advanced space travelers. But maybe time travelers or fallen angels; it is obvious they are deceivers who want to take over all our artificial intelligence and our Super-intelligence, thus controlling all our finances, economy, and visual and audio technology. This will enable them to control the world.

At first, they claimed they needed to meet with the people in person, but that was to prevent confusion caused by the news and media. They aim to rebuild our system to create a unified world with fewer influential people or our one true God in the mix. Things are happening at lightning speed; This is end-time stuff!

Walking into an
old-time country church,
which is overly crowded
with God-fearing people,
is a majestic beauty
that I shall never forget.
~ James L. Eakins

STAY READY

We all need to stay vigilant, as we are in a time when even a seemingly convincing lie can be believed. Fake and real can easily be confused, and God is not a God of confusion. As Christians, we understand the times and seasons, knowing that the day of the Lord will come like a thief in the night. Jesus Christ could return at any moment for his church. Therefore, we must gather around his preserved printed word.

The world is claiming its victims not only through graveyards but also through exciting, deceptive lies. However, we are not in darkness; this day will not overcome us. As a church awaiting the return of Jesus Christ, we must pray and walk in the Spirit alongside other faithful Christians.

Gathering as a church is no longer an option but a fundamental conviction. Do not seek spiritual nourishment and guidance from computers or media sources, as they will be compromised. In contrast, the printed word of God and the gathering of saints around the Bible, guided by the Holy Spirit, cannot be contaminated by the world, just as Jesus and his apostles had no computers, cell phones, or media. We can survive, too. If you choose to rely on your gadgets, that's your decision, but do not put too much trust in them.

Lord, help me never to be showy;
Just let me stand in your light.
~ James L. Eakins

PHONE CALL
FROM ATLANTA, GEORGIA

It was early Monday morning; I had just arrived in my church office and started preparing my week's schedule. My secretary Jude buzzed my phone and said, 'You have a call on line one.' A pastor named Doctor Lucas from Atlanta, Georgia, is insisting on speaking with you. Initially, I knew what he wanted me to do: attend his exceptional service, which was in honor of his special guest from another world. Picking up the call, I greeted Dr. Lucas with a joyful hello, and it's great to hear from you. How is everything going? He said, "Pastor James, I have your plane ticket and motel reservation all ready for you. The meeting is next Friday night, and I have a front row seat for you on my church platform. What else could I say but I'll be there.

The name of his church was LCCC, short for Logan Crockett Community Church. This church is one of the largest megachurches in Atlanta, Georgia, with a seating capacity of approximately 6,000 people. My reservations were at one of the finest hotels in Loganville, Georgia, just on the outskirts of Atlanta. Dr. Lucas spared no expense for me, and he even picked me up at the airport, providing me with a rental car. I made arrangements with Josh, our associate pastor, and reluctantly planned my trip.

Well, the grand day had arrived, and I flew into Hartsfield-Jackson Atlanta International Airport. Pastor Lucas and his wife, Judy, were there to greet me with hugs and smiles. We then went to a very fancy restaurant for a rib-eye steak. While eating, Lucas mentioned that my rental car was at the hotel, and it was a brand-new Shelby Mustang; he said, 'I hope you like it.' I laughed and said, "You've been talking to Josh, haven't you?" Lucas replied, "Actually, I've been listening to you preach on your church's YouTube." When I arrived at the hotel, I saw that the Shelby Mustang was a stunning cherry red.

Lucas and Judy dropped me off at the hotel and said, "We'll see you tonight. Please arrive an hour early, as I would like to introduce you to our honored guest. I thought to myself that really isn't one of my desires tonight; I didn't trust these beings from another world.

While driving to the meeting in my very nice rented Shelby Mustang, I was introduced to Officer Stopp. Yep, I got pulled over by a police officer named Stopp. He asked me if I knew why he had pulled me over. In reply, I said I suppose I was speeding, and by the way, is your name really Stopp? He didn't answer me. He asked me, "Where are you going in such a hurry?" I responded that I was going to the LCCC, where Dr. Lucas is the pastor. Officer Stopp said, "That is where I attend church. Pastor Lucas is a great man of God. I'll be directing traffic there in just a few minutes." He added, 'I'll tell you what, I'm going to let you go with a warning this time. Since you're in such an amazing hot

Phone Call From Atlanta, Georgia

rod car and in such a big hurry, and so am I, get behind my car and follow me.' Officer Stopp then turned on his flashing lights, and away we went to the church.

When I arrived at the church, it was already packed. Everyone wanted to meet these visitors from another world, or is it another dimension? One of the church elders came up to greet me by name. He said that Pastor Lucas had already told him that I would be arriving in a bright red Shelby Mustang. The elder said, 'My name is Bobby, and come with me.' He then took me through a secluded door, and as I walked in, I was surrounded by church offices, one of which belonged to Dr. Lucas. Bobby said, "Pastor Eakins, have a seat, and Pastor Lucas will be with you shortly. Lucas came quickly and instructed me on the upcoming activities. He informed me that there would be only two visitors, and they would be giving a presentation. However, Pastor Lucas said we will have a full-blown worship and praise service to God before these two visitors begin their presentation. Putting God first did make me much more at ease. The pastor then led me to the church's platform.

All the singers and musicians were in place and ready to worship and praise God. Then the two 8-foot-tall visitors came and towered over me. Lucas hurried to introduce me to them. I remembered one of them; his name is Lucian. Our last encounter was in Chicago, and it did not go too well. I'm sure Lucian knows I don't trust him. But Lucas was gracious and introduced me to both of them, one named Lucian and the other named Shofar. Lucas shared my name with them as well. Shofar made me nervous; he was always fishing for attention.

It's now time to start the service, and Pastor Lucas opened with a wonderful greeting, introducing the 8-foot-tall visitors, named Shofar and Lucian, and thanking them for coming to share some of their fantastic news. It was at this point in the introduction that the visitors started waving and smiling, and wanted to take over the service. Pastor Lucas said, "But first, let us worship and praise our one true God." Well, the two tall guys lost their grins. I had a sick feeling that this night wouldn't turn out well.

Then the music started, and it was loud and beautiful. The praising and worshiping were dynamic. However, the two visitors continued trying to get the people's attention. At times, they would hold their ears shut, and it looked like they were cursing. Some claimed they overheard them cursing. It was embarrassing to watch them vie for the spotlight of attention. Looking over at Pastor Lucas, I could tell he was extremely worried. Then it happened—something unexpected, something that had occurred in Chicago with a heckler there. Both of the strange beings raised their hands as if to say stop. The room fell silent; all the musical instruments were muted, and all sound in the church disappeared. But all the praise and worship to God continued loudly in a cappella, "a cappella." **Praise God, no one or nothing can silence God's worship and praise. These beings could not mute**

God's glory, and Pastor Lucas is seeing it all right here before his eyes. They could mute every sound in the sanctuary, but not the voices of worship and praise to God.

God Knows

Jesus knows our darkest thoughts.
He even knows our troubled hearts.

He knows what's behind our smile.
He loves us all the while.
He gives to us in his majestic style.

Don't give up and don't cave in.
God's love for you will always win.

~ James L. Eakins

THE SERVICE
WAS SUDDENLY CANCELED

Pastor Lucas then abruptly canceled the visitor's presentation. Some of the crowd left angry, but my brother Lucas did what was right. The two strange visitors vanished. They had a stormy night too!

Pastor Lucas came to me, trembling. He said, "How could I have been so deceived? Pastor James, you were right in Chicago. Lucas, trembling, asked me to come to his office. In private, Lucas cried and said, "I'm scared for my life." I told him that God wasn't going to hurt him; we all make serious mistakes sometimes. Lucas said, "I'm not afraid of God killing me. I'm fearful of these supernatural beings killing me." Lucas went on to tell me he had promised to promote and support these Visitors from another dimension.

Responding to Lucas, I suggested that perhaps you and Judy should take a sabbatical and disappear. You are welcome to visit our church and find rest. We will accommodate you for a period of two to three months. Just don't tell your board or church where you're going; leave your church in the good hands of your assistant pastors. Instruct them to have no guests and no special meetings, and that you will have a weekly conference call with them.

Make sure to encourage everyone not to trust the media or the internet, but instead focus on an old, printed copy of the Holy Bible. Tell the preachers to strongly urge the people to gather together, as uncertain times are approaching.

Lucas and his wife agreed to come to our church for a rest and to escape from these supernatural beings, as I was leaving for the night to rest in my hotel room. Before leaving, we had a heartfelt prayer, and I told him that as I fly out in the morning, I will return the Mustang to the rental station next to the airport. Driving away, I thought softly in my heart that Lucas and I have become special friends.

A BIG SURPRISE
ON MY WAY OUT OF ATLANTA

My fight was at five o'clock the next morning, so I was up at 3:00 am. I'm shaving, and my phone rings. It's Brother Lucas, and I'm thinking that this call has got to be bad news. Very nervously, I didn't answer with "hello", but are you alright? He laughed and said I've never been better. In all the confusion, I forgot to tell you that your plane flight was canceled yesterday. I said, "Now, Lucas, that is terrible news. How am I going to get home, and who canceled my flight?" Lucas laughed and said I canceled your flight. Then I asked with irritation, "Why did you cancel my flight? Lucas said, 'Drive your brand-new cherry red Shelby Mustang home, it's all yours, it's bought and paid for, and the taxes are paid too, your name is on the title, and in the mail. Before I left, I had lunch with Lucas and Judy and told them I couldn't wait to have them in our church. I also invited Lucas to preach in my pulpit. As I left for home, I told them I would see them in three days and that I was overwhelmed with such a great friendship.

Well, I'm now driving and reflecting on everything that has happened over the last three days, and amidst the roar of my Mustang, I heard the sound of a siren; it was Officer Stopp, pulling me over again. Approaching my car window, he asked, 'Do you know why I pulled you over? I asked if I was speeding. He said you're never going to learn to slow down, are you? But that's not why I pulled you over. He said, 'I want to take a closer look at your amazing car.' We chatted for a while, then exchanged our farewells.

AFTER GETTING HOME,
I EAGERLY AWAITED LUCAS AND JUDY'S ARRIVAL AT OUR CHURCH

Four days had passed, and still no Lucas. I called his cell phone several times, and I got no answer. I did not have Judy's number, and to contact the church under the cloak of silence would be a wasted effort.

It's Friday, and I'm in my office preparing for Sunday morning and Sunday night services. My secretary, Jude, buzzes me that there is a call on line two, and it is very urgent. Asking my secretary for more information, she said, "It's bad news from Logan Crocket Community Church, where Doctor Lucas is the pastor. Then, picking up the call, on the other end of the line was Bobby, who had first met me in the Church parking lot and ushered me into the church's office to meet Pastor Lucas. He was weeping. Barely getting the words out, he said, "Preacher, Doctor Lucas, and Judy are dead." It hit me like a thousand tons of bricks, a cold sweat broke out on me, and I was sick to my stomach. I prayed with Bobby, though I don't remember my prayer. I was in unbelievable shock. Hanging up, I just sat in sorrow and disbelief for another hour. I had sent my secretary home. I just wanted to be alone.

The next day, I called Pastor Lucas's church to get information on what had actually happened. They reported that it was a suspicious one-car accident that resulted in a destructive fire, leaving hardly any remains. Some of the things that happened to their car defy the laws of gravity. The police have not ruled out homicide, saying it looks more like an execution than an accident.

After this report, I can still hear Lucas saying, "I'm afraid that these supernatural beings are going to kill me." However, my friend Lucas made the right decision on that dark Friday night when he canceled the presentation of the two mysterious visitors from another dimension.

The church in Atlanta asked me to officiate the funeral service for Lucas and Judy, and I was honored to do so. But with it would come many tears.

THE TRIBUTE AND FUNERAL

After arriving in Atlanta, Police Officer Stopp was watching over the interstate. He pulled in front of me with all his patrol car lights flashing, motioning for me to follow him through the busy traffic. As soon as I reached the church parking lot, I thanked Officer Stopp and hurried into Pastor Lucas's megachurch. It was the day of Lucas and Judy's funeral service. I arrived two hours early, and Bobby met me in the parking lot to escort me to Lucas's personal church office. The last time I was here was on that late Friday night when Lucas confided that he feared for his life after abruptly canceling the visitor's presentation because he could no longer stand with these weird, strange beings. This office now feels cold and disappointing. However, I must finalize the previous arrangements I made for the funeral service by phone.

There would be a lot of music, including praise and worship, dedicated to God. Three special songs were to be "God, How Great Thou Art," "All Is Well with My Soul," and an old-time favorite, "When We All Get to Heaven." These songs would be woven in and out through the praise team's music. After the music, I would recognize the Governor of the great state of Georgia and all the other dignitaries. There would also be hundreds of ministers, both locally and from around the world.

The eulogy would be bittersweet, yet easy to express, because Lucas and Judy were great servants of God. It was unusual that they had no children or living siblings, but rather a large number of brothers and sisters in Christ.

Because Lucas and Judy influenced so many people around the world, we will be broadcasting and livestreaming their funeral service. It's now time for the service to commence. As I take my place on the large platform, there is hardly any room for the praise team. Thousands of beautiful flowers fill the space, and they certainly smell lovely. Looking over the crowd, it resembles a sea of faces; the sanctuary is jam-packed with people showing their love and respect for Lucas and Judy. However, towering over the crowd were two visitors, Shofar and Lucian. I was a little taken aback, but I'm sure they weren't here out of respect, but for other selfish reasons. They were still working to gain the people's gullible trust. I prayed that God would not let them disrupt or defile Lucas and Judy's service.

After all the greetings and music, I spoke for nearly forty minutes, mainly about the outstanding accomplishments and beautiful lives of Lucas and Judy. However, something weighed heavily on my heart, and I felt the need to speak out and preach a little as well.

I told them that our world is deeply deceived and changing at lightning speed. It has become increasingly difficult to distinguish between what is real and what is not, as well as between what is true and what is false.

The Tribute and Funeral

There is an old, enduring song we need to sing and live by, entitled "Hold to God's unchanging hand," and we should always hold to God's preserved, written word, the Bible.

Jesus Christ has declared that Heaven and Earth shall pass away, but His words shall never pass away. Jesus Christ is our only way off this planet, alive and forgiven. He was sacrificed, died for our sins, and conquered death, hell, and the grave. Let us never forget that God's word is forever settled in Heaven.

As I concluded my preaching, I addressed Christians everywhere: Do not trust Artificial Intelligence or Superintelligence for any spiritual guidance. Beware of computer-generated videos or printed materials. In fact, trust nothing except the Bible, God's written word. In these deceiving times, God's people must gather around the preserved Holy Bible, the written word of God. Do not rely on electronic devices for truth and spiritual enlightenment. Gather together in Jesus' name all the more today, as we see a world spiraling away from our true God and moral truth. Let us take a tangible Bible, an old written copy of God's word. Spread it before us and trust, believe, and worship Jesus Christ.

After concluding the service, the two visitors, Lucian and Shofar, towered over me and offered to shake my hand. I refused the gesture. They asked, "You don't trust us, do you?" I replied, "Trust is earned, not given." Then Shofar inquired, "How can we earn your trust?" I responded, "You can't. I don't entertain deception." Lucian then asked where my church was located. I replied, "We're not looking for visitors who are cunning deceivers." They hurried away, disappearing into the massive crowd; even with their tall stature, they seemed to vanish.

Returning home, Officer Stopp pulled me over with his lights flashing one last time. As always, he asked, 'Do you know why I pulled you over?' Of course, I did. I was speeding home. Officer Stopp said, 'I just wanted to tell you goodbye and Godspeed, my friend.' He went on to express his gratitude for the amazing tribute service in honor of Lucas and Judy, and mentioned that he appreciated my boldness about the Bible. He agreed that in these last deceptive times, the church must gather around God's written word. Before I let Officer Stopp go, I asked him if he thought Lucas and Judy had been murdered. He said, 'I believe some unknown powers executed them.' He told me that the homicide division was still investigating the case. Then, I asked him if he thought I might also be in danger. Officer Stopp replied, 'Why do you think I've been around you in your every move?' He added as I drove away, 'Don't be surprised if you see an unusual, continual police presence at your Ozark Gospel Church when you get home.'

Upon arriving home safely, I found our church secure in the sweet embrace of God and His Holy Written Word. Jesus is our Savior, and we shall not be moved. We are anchored in Jehovah. My faith in God has never been stronger, and I eagerly await the swift return of Jesus Christ.

THE CHURCH
CONTINUES TO GROW

With open doors, our church continued to experience rapid growth despite the deception surrounding us. Many were coming in search of answers and security during these perilous times. The beautiful truth that we have a Bible, which is both tangible and spiritual, and has stood the test of time, comforts them greatly. As a pastor, it was my responsibility to keep the people focused, believing, and trusting in God's Holy Word, despite the constant deception from the misled masses outside.

There were even pastors and churches that believed and preached that God was sending these visitors from another dimension, along with our inventions and superintelligence, to guide us into a new kingdom.

After a Sunday morning music and preaching service, a man approached me and asked if I remembered him. He mentioned he was from Dallas, Texas. It was Pastor Bobby Cain, whom I had first met along with Pastor Tommy Demas in Chicago, Illinois. It was there that I also met my good friend, Doctor Lucas. I then inquired about how his small church was holding up. He replied that their church had only about forty members and that the building was modest. We often had visitors attend our little service, spreading outrageous lies. They could be aggressive at times. After much prayer, I decided to move our church services into our homes. There, we would not be disturbed and could focus on the written word of God. He asked me how we managed order in our church. I told him it was with our solid members and a strong security team, strong preaching and prayers, and amazing worship and praise services, all under the firm hand of God, giving no place to the devil or his deception.

Then I asked Bobby about Pastor Tommy Demas, who lives in Dallas. Bobby began to cry and, with a choked voice, said that Tommy had been murdered. His church had around 400 attendees, and he had taken a stand against the rise of new deceptions spreading across the land. Tommy was shot in the head by a high-powered rifle through a window while he was preaching. The church is now deeply divided over the changing world, with some believing, like other churches, that these visitors from another dimension are men of renown sent to us by God.

After spending time with Pastor Bobby, I encouraged him in the Lord and urged him to keep a written Bible available for his congregation. However, it is crucial for all of us to be outspoken witnesses to the outside world; people need salvation and the God of hope in these times of sin and sorrow. Our faith in God and His written word must be stronger than ever.

It feels like every day brings a new disaster, some natural and some man-made. Just two nights ago,

The Church Continues to Grow

China was struck by a 9.5 magnitude earthquake. They are still counting the dead. Last week, Russia experienced a 9.0 earthquake along the Ring of Fire, triggering a tsunami warning that reached as far as Hawaii. Even now, Hawaii is covered in volcanic ash from a recent eruption, and the air is filled with smoke and toxins that cause eyes to burn and make breathing difficult. Massive mudslides in California have shut down miles of roads, destroyed many homes, and claimed hundreds of lives. Floods are causing fatalities worldwide, with the U.S. bearing the brunt of many EF5 tornadoes. Many hurricanes are Category Five.

Week after week, we see mass killings and cities burned to the ground by rioters. There's a rise in illness, mental health issues, and diseases, many of which have no cures. People are scared and confused. Over the past three days, we've been in total darkness. I don't mean the power is out; I mean there's no sun, moon, or stars visible. Sometimes, a ring of fire appears in the sky and then disappears. People often report strange flying objects overhead, and the media is flooded with bizarre and unusual videos. Many of these videos pose as trustworthy news, promoting visitors who have come to planet Earth.

The visitors from another dimension, possibly time travelers, are still somewhat active. They tend to appear in large gatherings, trying to guide the masses toward their idea of super artificial intelligence. Too often, they show up at a pastor's door when he is alone, convincing him that they are messengers from

God. Their sudden appearances and disappearances right before his eyes often seem to convince him they are sent to help in the world's recovery. I'm saddened to say that most churches are falling prey to this tactic. These visitors are no longer as active as they once were, but I have a bad feeling that they are meeting secretly not only with pastors but also with church leaders and world government officials.

Recently, world news confirmed that these visitors from another world posed no new threat to our planet despite their size and high intelligence. NATO, the Vatican, Congress, and the White House have all extended invitations for them to speak. They will discuss natural disasters, solutions to stop cyber hacking, and how to manage AI with its Artificial Superintelligence.

Things are changing rapidly. So, I keep telling our church to stay close to your written Bible because God's word can be trusted. The funny but sad part is that our church has been broken into and robbed several times, but they have never once stolen a Bible.

A CASHLESS SOCIETY

It was a Sunday morning that I will never forget. After the service, Brother Carl and Sister Sharon approached me, looking very concerned. Carl asked, "Pastor, how is our church going to survive in a cashless society?' I said I hadn't thought about it, and Carl replied, "You'd better start thinking about it and come up with a plan." We are going to be cashless very soon; it will all be digital money in the near future. I quickly went home and caught up on the news. Sure enough, the President and Congress had declared that coin and paper money, along with most paper checks, would no longer be recognized as legal tender. Everyone had just 90 days to convert coins and cash into their personal bank accounts or their government-issued debit cards. The government was offering a debit card with a $2000.00 balance already loaded onto it. They were informed that by faithfully using this government card, additional free money would be added to their debit card number from time to time. I strongly urged our church not to participate in this program, but I would understand if it were a complete necessity for some of them to have this card. What's scary for man's convenience is that very soon, a person will no longer need to carry a card or a device; a number can be put under the skin of the hand or forehead. **This will be a great big no for all of us!**

The coins and cash will become worthless after 90 days. From what I understand, the coins would be part of the material to make sidewalks and roads. The coins will be mixed into the rocks and concrete used to build the roads. They are going to throw a lot of silver into the streets. I suppose our coins are made from such a mix of materials that melting most of them down wouldn't be worthwhile.

Most of the stores had already stopped accepting cash, and many of their cashiers struggled to count money; without their computers, they were helpless. In fact, without their computers working, the store would become paralyzed and have to close. The banks despised the large coin deposits, and most of them stopped accepting coins. Thus, a large number of coins were donated to our church.

Our church still gathered every day with large crowds. Music, prayers, and preaching from God's written word remained our top priorities. We spent many hours each day with the crowds arriving. As we sought answers from God, a growing sense of restlessness spread among the people, as they sensed that something catastrophic was approaching.

THE LAST CASH FLOW

As strange as it sounds, cash started coming into our church by the bushels and coins by the buckets. It seemed that some of our older worshippers had saved large amounts of cash over the years and didn't want to put it into their bank accounts. They didn't see a need to do so at their age, so they just brought their cash and coins to the church. The church has had a debit and a credit card with our local bank for years. This money was coming into our church at a rapid rate. This was God providing for the upkeep and maintenance of the church in a way I hadn't expected. However, this blessing will all come to an end in 90 days.

Josh, our pastor and music minister, was a talented musician. As a result, some people gave him a large sum of cash. Therefore, Josh needed to deposit this cash into his bank account. He then planned to buy a custom-made Fender guitar for $29,000, of course, with the blessing of God's people.

The next day, Josh went to the music store to buy the Fender guitar, but his debit card was declined. He called the bank to verify it was really him making the purchase, but that wasn't the issue. The bank required a special government online form to be completed for purchases exceeding $7,000.00. Josh was furious; of course, the politicians and the elite have no such requirements. But now everyone could see how this buying and selling was going to play out in the last days. Josh finally purchased his custom-made Fender guitar, a fantastic instrument. The guitar sounds amazing during our music and praise sessions as we have continuous services and stay focused on the written word of God.

    Well, I'm at the church. It's 3 o'clock on Sunday morning, and while I'm praying alone, two visitors suddenly appear before me. They didn't come through the church doors; they just materialized in front of me. It wasn't Lucian or Shofar, but two other extremely tall and muscular men. They introduced themselves as visitors from the ancient promised land of Abraham. One was named Iscariott, and the other's name was Anakim. Immediately, I asked, "Do you know anything about the mysterious death of Lucas and Judy, who were servants of God in Atlanta, Georgia?" They replied, "This is not in our liberty to discuss." Then I asked if they knew Lucian and Shofar. They said they knew them well and that Lucian and Shofar are assigned to the larger cities, while they have been assigned to smaller rural areas and towns. Who assigned you? I asked. "They said, we can only tell you that the watchers of Earth have a new plan for your world." They responded with eagerness to discuss the watchers' plans. They informed me that the watchers have bestowed upon the Earth great leaders, paving the way for a new and better world, and advised me to encourage the people to follow these new leaders and adhere to their programs.

One of the visitors, named Iscariott, asked if they could speak with our Sunday crowds, and he strongly specified that there be no music or preaching around them. That's when I demanded in Jesus' name and Christ's authority that they must leave. Around that time, we heard the front door of the church open, and it was Brother Jimmy coming in. He had been warned in a dream that something was wrong at the church and his pastor was in danger. The two visitors became very nervous and suddenly disappeared from my sight.

Jimmy didn't notice the two visitors, but as he drove into the church parking lot, he saw a ring of fire in the dark, cloudy sky above the church. The ring of fire disappeared as he opened the church door. Jimmy told me all about his dream and the ring of fire. He also mentioned an eerie feeling in the church when he first arrived. We spent the rest of the early morning hours in praise, worship, and prayer. After this encounter, I noticed a heavy police presence around the church, and the officers would follow me to and from the church, mainly because of the destruction and death threats against our church.

Let not your heart be troubled:
ye believe in God,
believe also in me.

In my Father's house
are many mansions:
if it were not so,
I would have told you.
I go to prepare a place for you.

And if I go and
prepare a place for you,
I will come again,
and receive you unto myself;
that where I am,
there ye may be also.

By Jesus Christ

THE CROWDS KEEP COMING
TO OUR CHURCH FOR ANSWERS

Week after week, our church team continues to offer spiritual and genuine worship that is truly godly. As I continue to preach and teach from God's written word, we are strengthening our faith in God and His established promises more than ever today. In every service, I remind people that Jesus Christ will return soon; don't be deceived by trusting others. We, the church, will be taken up at Jesus Christ's return and will enjoy the safety of eternal life.

Although the crowds have been enormous, I'm beginning to notice a change in people; they are starting to feel more comfortable with their economy and are living a superficial life filled with pleasures around them. It seems the fear of God's judgment and the return of Jesus Christ is fading from their minds.

My message now is that the church will suddenly be caught up to meet Jesus Christ, and then God's wrath will begin on Earth. Occasionally, the two tall visitors, Iscariott and Anakim, will appear and disappear before or after the music and preaching. Still, Josh and I are the only ones who can see them. Jimmy always misses out on seeing them appear and vanish again and again.

It's been three exciting years now, but some of the

people are falling back into a state of lukewarmness. Our crowds of visitors are becoming smaller, but we still have a large gathering of Christian believers daily around God's written word. **Our faith has never been stronger, and our church is filled with true believers.**

Due to financial restrictions on our digital money, we're only allowed to spend small amounts and can only contribute a limited sum to the church. Our overseers must approve major purchases; of course, the wealthy do as they please. Because of this, our members barter, exchanging goods or services without using money. Mega churches are also succeeding in this regard. In our church, different members take turns paying the bills with their debit cards.

BIG D' STARTS COMING
TO CHURCH

It's Sunday night, and we're experiencing a fantastic service. Gary is tickling the ivory on the grand piano, and Jimmy is moving swiftly on the drums. The crowd is raptured in worship. Heaven's glory cloud was hanging low, filled with abundant blessings. We all rejoiced that Jesus was soon to return, and we would be caught up in the glory clouds to meet Him in the air; and so, shall we ever be with the Lord.

After the service, a big and tall man approached me and introduced himself as Billy Dee. He said, "But everyone calls me Big-D." I replied, 'Big-D, it's a privilege to meet you." Then Big-D told me he loved the service and was deeply moved in his heart by the peace, love, and joy in the crowd. But he asked with a smirk, 'You really don't believe that a man named Jesus is going to snatch all of you off this planet in the twinkling of an eye, do you?' I shouted, 'Yes, I do.' Around that time, I took a large Bible and opened it to Matthew 24, then placed it on top of the pulpit. Big-D asked, 'What's this all about?' I replied that every time after the service, I place an open Bible on the pulpit just in case the church is caught away. So, when people fill the church, and I won't be here, God's written word will still be open to them in this troubled world. Big-D said, 'I don't believe it, but I love the atmosphere in this place.' I told him, 'Big-D, you'll come around, you just stick around, and you'll see for yourself."

Big-D asked, 'Do you know what the Big-D stands for?' He said it stands for big dynamite. I replied, "Do you know what Big-D means to the child of God? It means the Devil." Well, Big-D said with an inflated chest, 'My dynamite will blow the devil to smithereens.' I got up in Big-D's face and said, 'The Holy Ghost is the only dynamite that can blow the devil to smithereens." Big-D got up in my face and said, 'I like you, preacher, I like you.' As Big-D and I walked away from the pulpit, Big-D turned back and looked up at the open Bible left on the pulpit. He said, **'Preacher, that's a good idea just in case we are gone."**

One early Sunday morning, I was praying at the altar. It was 4 o'clock, and still dark outside. Someone was pounding on the locked front door. I knew it wasn't one of the time travelers; they never use a door. As I made my way to the door, I saw it was Big-D; he was excited, to put it mildly. The second he walked in, he said, 'Preacher, two of those strange giants broke into my house just a little while ago, and I didn't even let them talk. I told them the only person I want to talk to is Jesus Christ, and they had nothing I wanted to hear.' Then I told them to get out of my house, and they just vanished right before my eyes. I said Big-D, tell me more about it. He said nope, they're just doing the devil's work. Preacher, I'm really here to share with you a great Idea.

BUCKETS OF COINS

I said, 'Okay, Big D, I'm all ears.' Please tell me your great idea. He said he met Chuck at church last Sunday — you know, the Chuck who is a professional stone and concrete artist. I think his dad is Curtis. I've seen some of his work, and it's spectacular. You know all the coins the church has in the back room? I replied yes, and I don't know what to do with all 199 five-gallon buckets full of them — filled with pennies, nickels, dimes, quarters, and half dollars. Big D says, "Here's my brilliant idea: let's get Chuck to create us a sidewalk of coins, we can call it 'The walk of change.'" I said, "Big D, that is a brilliant idea. We can also lay one of our classroom floors in coins and call it "Room for Change."

Well, Chuck got right on it, and the sidewalk full of coins was created; the classroom floor and sidewalk were decorated with shiny coins. We also finished two-bathroom floors with shiny coins. We call them our 'Shiny Privy Rooms." People came from all around to walk on The Walk of Change, and some of our members were always ready with an open Bible to share the gospel and pray for those in desperate need.

Always remember that
your personal salvation testimony
is more powerful
than you can ever imagine.

*And they overcame him
by the blood of the Lamb,
and by the word of their testimony;
and they loved not their lives
unto the death.*

(Revelation 12:11)

THERE IS ANOTHER GREAT SHAKING
COMING UPON THE EARTH.

Over time, people have become numb to fearing the end times and God's judgment. Israel currently enjoys peace, though rumors of war still spread everywhere. Meanwhile, North Korea makes threats toward Israel. (I'm surprised by Israel's temporary peace.) Still, terrible tragedies continue to occur around the world. Just a few nights ago, one of our faithful church members rushed into the church, sobbing and trembling, describing a giant asteroid that had just hit California, causing many casualties and widespread damage in California and Arizona.

Another upheaval is approaching the land. It could be a nuclear explosion or more visitors from another dimension. Maybe it will be the Rapture and the disappearance of the faithful saints in Christ. Week after week and day after day, I preach the sudden return of Jesus Christ and urge people everywhere to repent of their sins, stay focused on Jesus the Son of God, and turn to God's written Word. Because there is no doubt that when the trumpet of God sounds and the church is taken up to meet their Lord in His glorious cloud, millions of people will be left behind to face God's great judgment, this departure of the church will cause a great shaking upon the earth, leading to the church buildings and altars filling with unbelievable crowds. Many lost people will cry in and at the altars of the church. They will be empty, fearful, and shaken in their hearts and minds. You may be one of these left behind, but you can still be saved in this terrible time.

Why not give your all to Jesus Christ today and trust in Him for your redemption? For whosoever calls on the name of the Lord Jesus Christ shall be saved. Remember, Jesus shed His blood on the cross for your forgiveness, died for your punishment of sin, and rose from the grave to give you everlasting life.

For God so loved the world that he gave his only begotten Son that whosoever believeth in him should not perish but have everlasting life.

Soon, the true church of the living God will be taken from the earth just before the great and terrible Tribulation. God's wrath is approaching! But God has not destined His church for His wrath. **One day, the true church will vanish, and God's wrath will commence shortly.** To you, my faithful reader, please prepare and don't be left behind.

WHAT IF VISITORS DO ARRIVE
FROM ANOTHER PLANET OR DIMENSION?

If you woke up in the morning to the news that we had visitors from another planet, **what would it do to your faith?** For me, the answer is both exciting and wonderful. I know God created it all, and **Jesus is Lord**.

These visitors would only show that God is greater and more powerful than we have ever imagined.

God so loved the world that he gave his only begotten Son that whosoever believeth in him should not perish but have everlasting life.

The wages of sin is death, but the gift of God is eternal life through Jesus Christ our Lord.

If you are afraid of dying in sin and your life matters greatly to you, as well as where you will go after death, Jesus is the only way, the truth, and eternal life. From your heart, pray this prayer to God.

Dear Lord Jesus, I acknowledge that I am a sinner, and I humbly ask for Your forgiveness. I believe that you died on the cross for my sins and rose again from the dead. I am turning from my sins and invite You into my heart and life. I completely trust you as my Lord and Savior. Amen.

The prayer you just offered emphasizes several key elements:

__Acknowledge sin:__ Recognizing that we have strayed from God's Holiness and need forgiveness.

Faith in Jesus: Believing that Jesus is the Son of God who died for our sins and was raised from the dead.

Surrender means: inviting Jesus to take complete control of our lives and trusting him as our Lord and Savior.

Repentance: God has opened your heart to turn away from sin and pursue forgiveness and a new life in Christ. We grieve away from our sin into the strong, loving arms of Jesus Christ.

AFTERWORD

Our world faces challenging times. The merging of sonic technology and superintelligence could lead to great suffering. Humanity may have created an unavoidable problem that traps us and limits our freedom to act. Perhaps our most incredible creation has become uncontrollable. It will require divine intervention. This book has presented a possible scenario in which our world could collapse. God must guide His children through His Written Word.

Pastor James L. Eakins

COMING QUOTES
OF THE BEAST

We can make peace through war.

Let us eliminate all divisions among us.

We must save our world.

We are the world to come.

One great political leader for all.

One great religious leader for all.

One world order.

We are the world.

*If you would like to contact the author,
please write to:*
James L. Eakins
1302 South 18th Avenue
Ozark, Missouri 65721

MATTHEW | CHAPTER 24:35-43
King James Version

35 Heaven and earth shall pass away, but my words shall not pass away.

36 But of that day and hour knoweth no man, no, not the angels of heaven, but my Father only.

37 But as the days of Noah were, so shall also the coming of the Son of man be.

38 For as in the days that were before the flood they were eating and drinking, marrying and giving in marriage, until the day that Noe entered into the ark,

39 And knew not until the flood came, and took them all away; so shall also the coming of the Son of man be.

40 Then shall two be in the field; the one shall be taken, and the other left.

41 Two women shall be grinding at the mill; the one shall be taken, and the other left.

42 Watch therefore: for ye know not what hour your Lord doth come.

43 But know this, that if the goodman of the house had known in what watch the thief would come, he would have watched, and would not have suffered his house to be broken up.

1 THESSALONIANS | CHAPTER 1:10
King James Version

¹⁰ And to wait for his Son from heaven, whom he raised from the dead, even Jesus, which delivered us from the wrath to come.

1 THESSALONIANS | CHAPTER 4:13-18
King James Version

¹³ But I would not have you to be ignorant, brethren, concerning them which are asleep, that ye sorrow not, even as others which have no hope.

¹⁴ For if we believe that Jesus died and rose again, even so them also which sleep in Jesus will God bring with him.

¹⁵ For this we say unto you by the word of the Lord, that we which are alive and remain unto the coming of the Lord shall not prevent them which are asleep.

¹⁶ For the Lord himself shall descend from heaven with a shout, with the voice of the archangel, and with the trump of God: and the dead in Christ shall rise first:

¹⁷ Then we which are alive and remain shall be caught up together with them in the clouds, to meet the Lord in the air: and so shall we ever be with the Lord.

¹⁸ Wherefore comfort one another with these words.

1 THESSALONIANS | CHAPTER 5:1-10
King James Version

But of the times and the seasons, brethren, ye have no need that I write unto you.

2 For yourselves know perfectly that the day of the Lord so cometh as a thief in the night.

3 For when they shall say, Peace and safety; then sudden destruction cometh upon them, as travail upon a woman with child; and they shall not escape.

4 But ye, brethren, are not in darkness, that that day should overtake you as a thief.

5 Ye are all the children of light, and the children of the day: we are not of the night, nor of darkness.

6 Therefore let us not sleep, as do others; but let us watch and be sober.

7 For they that sleep sleep in the night; and they that be drunken are drunken in the night.

8 But let us, who are of the day, be sober, putting on the breastplate of faith and love; and for an helmet, the hope of salvation.

9 For God hath not appointed us to wrath, but to obtain salvation by our Lord Jesus Christ,

10 Who died for us, that, whether we wake or sleep, we should live together with him.

1 CORINTHIANS | CHAPTER 15:51-52
King James Version

⁵¹ Behold, I shew you a mystery; We shall not all sleep, but we shall all be changed,

⁵² In a moment, in the twinkling of an eye, at the last trump: for the trumpet shall sound, and the dead shall be raised incorruptible, and we shall be changed.

2 THESSALONIANS | CHAPTER 2:1-12
King James Version

Now we beseech you, brethren, by the coming of our Lord Jesus Christ, and by our gathering together unto him,

² That ye be not soon shaken in mind, or be troubled, neither by spirit, nor by word, nor by letter as from us, as that the day of Christ is at hand.

³ Let no man deceive you by any means: for that day shall not come, except there come a falling away first, and that man of sin be revealed, the son of perdition;

⁴ Who opposeth and exalteth himself above all that is called God, or that is worshipped; so that he as God sitteth in the temple of God, shewing himself that he is God.

⁵ Remember ye not, that, when I was yet with you, I told you these things?

⁶ And now ye know what withholdeth that he might be revealed in his time.

7 For the mystery of iniquity doth already work: only he who now letteth will let, until he be taken out of the way.

8 And then shall that Wicked be revealed, whom the Lord shall consume with the spirit of his mouth, and shall destroy with the brightness of his coming:

9 Even him, whose coming is after the working of Satan with all power and signs and lying wonders,

10 And with all deceivableness of unrighteousness in them that perish; because they received not the love of the truth, that they might be saved.

11 And for this cause God shall send them strong delusion, that they should believe a lie:

12 That they all might be damned who believed not the truth, but had pleasure in unrighteousness.

THE BOOK OF REVELATION | CHAPTERS 1-22
King James Version

Revelation Chapter 1:1-20

The Revelation of Jesus Christ, which God gave unto him, to shew unto his servants things which must shortly come to pass; and he sent and signified it by his angel unto his servant John:

2 Who bare record of the word of God, and of the testimony of Jesus Christ, and of all things that he saw.

3 Blessed is he that readeth, and they that hear the words of this prophecy, and keep those things which are written therein: for the time is at hand.

4 John to the seven churches which are in Asia: Grace be unto you, and peace, from him which is, and which was, and which is to come; and from the seven Spirits which are before his throne;

5 And from Jesus Christ, who is the faithful witness, and the first begotten of the dead, and the prince of the kings of the earth. Unto him that loved us, and washed us from our sins in his own blood,

6 And hath made us kings and priests unto God and his Father; to him be glory and dominion for ever and ever. Amen.

7 Behold, he cometh with clouds; and every eye shall see him, and they also which pierced him: and all kindreds of the earth shall wail because of him. Even so, Amen.

⁸ I am Alpha and Omega, the beginning and the ending, saith the Lord, which is, and which was, and which is to come, the Almighty.

⁹ I John, who also am your brother, and companion in tribulation, and in the kingdom and patience of Jesus Christ, was in the isle that is called Patmos, for the word of God, and for the testimony of Jesus Christ.

¹⁰ I was in the Spirit on the Lord's day, and heard behind me a great voice, as of a trumpet,

¹¹ Saying, I am Alpha and Omega, the first and the last: and, What thou seest, write in a book, and send it unto the seven churches which are in Asia; unto Ephesus, and unto Smyrna, and unto Pergamos, and unto Thyatira, and unto Sardis, and unto Philadelphia, and unto Laodicea.

¹² And I turned to see the voice that spake with me. And being turned, I saw seven golden candlesticks;

¹³ And in the midst of the seven candlesticks one like unto the Son of man, clothed with a garment down to the foot, and girt about the paps with a golden girdle.

¹⁴ His head and his hairs were white like wool, as white as snow; and his eyes were as a flame of fire;

¹⁵ And his feet like unto fine brass, as if they burned in a furnace; and his voice as the sound of many waters.

¹⁶ And he had in his right hand seven stars: and out of his mouth went a sharp twoedged sword: and his countenance was as the sun shineth in his strength.

¹⁷ And when I saw him, I fell at his feet as dead. And he laid his right hand upon me, saying unto me, Fear not; I am the first and the last:

¹⁸ I am he that liveth, and was dead; and, behold, I am alive for evermore, Amen; and have the keys of hell and of death.

¹⁹ Write the things which thou hast seen, and the things which are, and the things which shall be hereafter;

²⁰ The mystery of the seven stars which thou sawest in my right hand, and the seven golden candlesticks. The seven stars are the angels of the seven churches: and the seven candlesticks which thou sawest are the seven churches.

Revelation Chapter 2:1-29

Unto the angel of the church of Ephesus write; These things saith he that holdeth the seven stars in his right hand, who walketh in the midst of the seven golden candlesticks;

² I know thy works, and thy labour, and thy patience, and how thou canst not bear them which are evil: and thou hast tried them which say they are apostles, and are not, and hast found them liars:

³ And hast borne, and hast patience, and for my name's sake hast laboured, and hast not fainted.

⁴ Nevertheless I have somewhat against thee, because thou hast left thy first love.

5 Remember therefore from whence thou art fallen, and repent, and do the first works; or else I will come unto thee quickly, and will remove thy candlestick out of his place, except thou repent.

6 But this thou hast, that thou hatest the deeds of the Nicolaitanes, which I also hate.

7 He that hath an ear, let him hear what the Spirit saith unto the churches; To him that overcometh will I give to eat of the tree of life, which is in the midst of the paradise of God.

8 And unto the angel of the church in Smyrna write; These things saith the first and the last, which was dead, and is alive;

9 I know thy works, and tribulation, and poverty, (but thou art rich) and I know the blasphemy of them which say they are Jews, and are not, but are the synagogue of Satan.

10 Fear none of those things which thou shalt suffer: behold, the devil shall cast some of you into prison, that ye may be tried; and ye shall have tribulation ten days: be thou faithful unto death, and I will give thee a crown of life.

11 He that hath an ear, let him hear what the Spirit saith unto the churches; He that overcometh shall not be hurt of the second death.

¹² And to the angel of the church in Pergamos write; These things saith he which hath the sharp sword with two edges;

¹³ I know thy works, and where thou dwellest, even where Satan's seat is: and thou holdest fast my name, and hast not denied my faith, even in those days wherein Antipas was my faithful martyr, who was slain among you, where Satan dwelleth.

¹⁴ But I have a few things against thee, because thou hast there them that hold the doctrine of Balaam, who taught Balac to cast a stumbling block before the children of Israel, to eat things sacrificed unto idols, and to commit fornication.

¹⁵ So hast thou also them that hold the doctrine of the Nicolaitanes, which thing I hate.

¹⁶ Repent; or else I will come unto thee quickly, and will fight against them with the sword of my mouth.

¹⁷ He that hath an ear, let him hear what the Spirit saith unto the churches; To him that overcometh will I give to eat of the hidden manna, and will give him a white stone, and in the stone a new name written, which no man knoweth saving he that receiveth it.

¹⁸ And unto the angel of the church in Thyatira write; These things saith the Son of God, who hath his eyes like unto a flame of fire, and his feet are like fine brass;

¹⁹ I know thy works, and charity, and service, and faith, and thy patience, and thy works; and the last to be more than the first.

20 Notwithstanding I have a few things against thee, because thou sufferest that woman Jezebel, which calleth herself a prophetess, to teach and to seduce my servants to commit fornication, and to eat things sacrificed unto idols.

21 And I gave her space to repent of her fornication; and she repented not.

22 Behold, I will cast her into a bed, and them that commit adultery with her into great tribulation, except they repent of their deeds.

23 And I will kill her children with death; and all the churches shall know that I am he which searcheth the reins and hearts: and I will give unto every one of you according to your works.

24 But unto you I say, and unto the rest in Thyatira, as many as have not this doctrine, and which have not known the depths of Satan, as they speak; I will put upon you none other burden.

25 But that which ye have already hold fast till I come.

26 And he that overcometh, and keepeth my works unto the end, to him will I give power over the nations:

27 And he shall rule them with a rod of iron; as the vessels of a potter shall they be broken to shivers: even as I received of my Father.

28 And I will give him the morning star.

29 He that hath an ear, let him hear what the Spirit saith unto the churches.

Revelation Chapter 3:1-22

And unto the angel of the church in Sardis write; These things saith he that hath the seven Spirits of God, and the seven stars; I know thy works, that thou hast a name that thou livest, and art dead.

2 Be watchful, and strengthen the things which remain, that are ready to die: for I have not found thy works perfect before God.

3 Remember therefore how thou hast received and heard, and hold fast, and repent. If therefore thou shalt not watch, I will come on thee as a thief, and thou shalt not know what hour I will come upon thee.

4 Thou hast a few names even in Sardis which have not defiled their garments; and they shall walk with me in white: for they are worthy.

5 He that overcometh, the same shall be clothed in white raiment; and I will not blot out his name out of the book of life, but I will confess his name before my Father, and before his angels.

6 He that hath an ear, let him hear what the Spirit saith unto the churches.

7 And to the angel of the church in Philadelphia write; These things saith he that is holy, he that is true, he that hath the key of David, he that openeth, and no man shutteth; and shutteth, and no man openeth;

8 I know thy works: behold, I have set before thee an open door, and no man can shut it: for thou hast a little strength, and hast kept my word, and hast not denied my name.

9 Behold, I will make them of the synagogue of Satan, which say they are Jews, and are not, but do lie; behold, I will make them to come and worship before thy feet, and to know that I have loved thee.

10 Because thou hast kept the word of my patience, I also will keep thee from the hour of temptation, which shall come upon all the world, to try them that dwell upon the earth.

11 Behold, I come quickly: hold that fast which thou hast, that no man take thy crown.

12 Him that overcometh will I make a pillar in the temple of my God, and he shall go no more out: and I will write upon him the name of my God, and the name of the city of my God, which is new Jerusalem, which cometh down out of heaven from my God: and I will write upon him my new name.

13 He that hath an ear, let him hear what the Spirit saith unto the churches.

14 And unto the angel of the church of the Laodiceans write; These things saith the Amen, the faithful and true witness, the beginning of the creation of God;

15 I know thy works, that thou art neither cold nor hot: I would thou wert cold or hot.

¹⁶ So then because thou art lukewarm, and neither cold nor hot, I will spue thee out of my mouth.

¹⁷ Because thou sayest, I am rich, and increased with goods, and have need of nothing; and knowest not that thou art wretched, and miserable, and poor, and blind, and naked:

¹⁸ I counsel thee to buy of me gold tried in the fire, that thou mayest be rich; and white raiment, that thou mayest be clothed, and that the shame of thy nakedness do not appear; and anoint thine eyes with eyesalve, that thou mayest see.

¹⁹ As many as I love, I rebuke and chasten: be zealous therefore, and repent.

²⁰ Behold, I stand at the door, and knock: if any man hear my voice, and open the door, I will come in to him, and will sup with him, and he with me.

²¹ To him that overcometh will I grant to sit with me in my throne, even as I also overcame, and am set down with my Father in his throne.

²² He that hath an ear, let him hear what the Spirit saith unto the churches.

Revelation Chapter 4:1-11

After this I looked, and, behold, a door was opened in heaven: and the first voice which I heard was as it were of a trumpet talking with me; which said, Come up hither, and I will shew thee things which must be hereafter.

2 And immediately I was in the spirit: and, behold, a throne was set in heaven, and one sat on the throne.

3 And he that sat was to look upon like a jasper and a sardine stone: and there was a rainbow round about the throne, in sight like unto an emerald.

4 And round about the throne were four and twenty seats: and upon the seats I saw four and twenty elders sitting, clothed in white raiment; and they had on their heads crowns of gold.

5 And out of the throne proceeded lightnings and thunderings and voices: and there were seven lamps of fire burning before the throne, which are the seven Spirits of God.

6 And before the throne there was a sea of glass like unto crystal: and in the midst of the throne, and round about the throne, were four beasts full of eyes before and behind.

7 And the first beast was like a lion, and the second beast like a calf, and the third beast had a face as a man, and the fourth beast was like a flying eagle.

8 And the four beasts had each of them six wings about him; and they were full of eyes within: and they rest not day and night, saying, Holy, holy, holy, Lord God Almighty, which was, and is, and is to come.

9 And when those beasts give glory and honour and thanks to him that sat on the throne, who liveth for ever and ever,

¹⁰ The four and twenty elders fall down before him that sat on the throne, and worship him that liveth for ever and ever, and cast their crowns before the throne, saying,

¹¹ Thou art worthy, O Lord, to receive glory and honour and power: for thou hast created all things, and for thy pleasure they are and were created.

Revelation Chapter 5:1-14

And I saw in the right hand of him that sat on the throne a book written within and on the backside, sealed with seven seals.

² And I saw a strong angel proclaiming with a loud voice, Who is worthy to open the book, and to loose the seals thereof?

³ And no man in heaven, nor in earth, neither under the earth, was able to open the book, neither to look thereon.

⁴ And I wept much, because no man was found worthy to open and to read the book, neither to look thereon.

⁵ And one of the elders saith unto me, Weep not: behold, the Lion of the tribe of Judah, the Root of David, hath prevailed to open the book, and to loose the seven seals thereof.

6 And I beheld, and, lo, in the midst of the throne and of the four beasts, and in the midst of the elders, stood a Lamb as it had been slain, having seven horns and seven eyes, which are the seven Spirits of God sent forth into all the earth.

7 And he came and took the book out of the right hand of him that sat upon the throne.

8 And when he had taken the book, the four beasts and four and twenty elders fell down before the Lamb, having every one of them harps, and golden vials full of odours, which are the prayers of saints.

9 And they sung a new song, saying, Thou art worthy to take the book, and to open the seals thereof: for thou wast slain, and hast redeemed us to God by thy blood out of every kindred, and tongue, and people, and nation;

10 And hast made us unto our God kings and priests: and we shall reign on the earth.

11 And I beheld, and I heard the voice of many angels round about the throne and the beasts and the elders: and the number of them was ten thousand times ten thousand, and thousands of thousands;

12 Saying with a loud voice, Worthy is the Lamb that was slain to receive power, and riches, and wisdom, and strength, and honour, and glory, and blessing.

¹³ And every creature which is in heaven, and on the earth, and under the earth, and such as are in the sea, and all that are in them, heard I saying, Blessing, and honour, and glory, and power, be unto him that sitteth upon the throne, and unto the Lamb for ever and ever.

¹⁴ And the four beasts said, Amen. And the four and twenty elders fell down and worshipped him that liveth for ever and ever.

Revelation Chapter 6:1-17

And I saw when the Lamb opened one of the seals, and I heard, as it were the noise of thunder, one of the four beasts saying, Come and see.

² And I saw, and behold a white horse: and he that sat on him had a bow; and a crown was given unto him: and he went forth conquering, and to conquer.

³ And when he had opened the second seal, I heard the second beast say, Come and see.

⁴ And there went out another horse that was red: and power was given to him that sat thereon to take peace from the earth, and that they should kill one another: and there was given unto him a great sword.

⁵ And when he had opened the third seal, I heard the third beast say, Come and see. And I beheld, and lo a black horse; and he that sat on him had a pair of balances in his hand.

⁶ And I heard a voice in the midst of the four beasts say, A measure of wheat for a penny, and three measures of barley for a penny; and see thou hurt not the oil and the wine.

⁷ And when he had opened the fourth seal, I heard the voice of the fourth beast say, Come and see.

⁸ And I looked, and behold a pale horse: and his name that sat on him was Death, and Hell followed with him. And power was given unto them over the fourth part of the earth, to kill with sword, and with hunger, and with death, and with the beasts of the earth.

⁹ And when he had opened the fifth seal, I saw under the altar the souls of them that were slain for the word of God, and for the testimony which they held:

¹⁰ And they cried with a loud voice, saying, How long, O Lord, holy and true, dost thou not judge and avenge our blood on them that dwell on the earth?

¹¹ And white robes were given unto every one of them; and it was said unto them, that they should rest yet for a little season, until their fellowservants also and their brethren, that should be killed as they were, should be fulfilled.

¹² And I beheld when he had opened the sixth seal, and, lo, there was a great earthquake; and the sun became black as sackcloth of hair, and the moon became as blood;

¹³ And the stars of heaven fell unto the earth, even as a fig tree casteth her untimely figs, when she is shaken of a mighty wind.

¹⁴ And the heaven departed as a scroll when it is rolled together; and every mountain and island were moved out of their places.

¹⁵ And the kings of the earth, and the great men, and the rich men, and the chief captains, and the mighty men, and every bondman, and every free man, hid themselves in the dens and in the rocks of the mountains;

¹⁶ And said to the mountains and rocks, Fall on us, and hide us from the face of him that sitteth on the throne, and from the wrath of the Lamb:

¹⁷ For the great day of his wrath is come; and who shall be able to stand?

Revelation Chapter 7:1-17

And after these things I saw four angels standing on the four corners of the earth, holding the four winds of the earth, that the wind should not blow on the earth, nor on the sea, nor on any tree.

² And I saw another angel ascending from the east, having the seal of the living God: and he cried with a loud voice to the four angels, to whom it was given to hurt the earth and the sea,

3 Saying, Hurt not the earth, neither the sea, nor the trees, till we have sealed the servants of our God in their foreheads.

4 And I heard the number of them which were sealed: and there were sealed an hundred and forty and four thousand of all the tribes of the children of Israel.

5 Of the tribe of Juda were sealed twelve thousand. Of the tribe of Reuben were sealed twelve thousand. Of the tribe of Gad were sealed twelve thousand.

6 Of the tribe of Aser were sealed twelve thousand. Of the tribe of Nephthalim were sealed twelve thousand. Of the tribe of Manasses were sealed twelve thousand.

7 Of the tribe of Simeon were sealed twelve thousand. Of the tribe of Levi were sealed twelve thousand. Of the tribe of Issachar were sealed twelve thousand.

8 Of the tribe of Zabulon were sealed twelve thousand. Of the tribe of Joseph were sealed twelve thousand. Of the tribe of Benjamin were sealed twelve thousand.

9 After this I beheld, and, lo, a great multitude, which no man could number, of all nations, and kindreds, and people, and tongues, stood before the throne, and before the Lamb, clothed with white robes, and palms in their hands;

10 And cried with a loud voice, saying, Salvation to our God which sitteth upon the throne, and unto the Lamb.

¹¹ And all the angels stood round about the throne, and about the elders and the four beasts, and fell before the throne on their faces, and worshipped God,

¹² Saying, Amen: Blessing, and glory, and wisdom, and thanksgiving, and honour, and power, and might, be unto our God for ever and ever. Amen.

¹³ And one of the elders answered, saying unto me, What are these which are arrayed in white robes? and whence came they?

¹⁴ And I said unto him, Sir, thou knowest. And he said to me, These are they which came out of great tribulation, and have washed their robes, and made them white in the blood of the Lamb.

¹⁵ Therefore are they before the throne of God, and serve him day and night in his temple: and he that sitteth on the throne shall dwell among them.

¹⁶ They shall hunger no more, neither thirst any more; neither shall the sun light on them, nor any heat.

¹⁷ For the Lamb which is in the midst of the throne shall feed them, and shall lead them unto living fountains of waters: and God shall wipe away all tears from their eyes.

Revelation Chapter 8:1-13

And when he had opened the seventh seal, there was silence in heaven about the space of half an hour.

² And I saw the seven angels which stood before God; and to them were given seven trumpets.

3 And another angel came and stood at the altar, having a golden censer; and there was given unto him much incense, that he should offer it with the prayers of all saints upon the golden altar which was before the throne.

4 And the smoke of the incense, which came with the prayers of the saints, ascended up before God out of the angel's hand.

5 And the angel took the censer, and filled it with fire of the altar, and cast it into the earth: and there were voices, and thunderings, and lightnings, and an earthquake.

6 And the seven angels which had the seven trumpets prepared themselves to sound.

7 The first angel sounded, and there followed hail and fire mingled with blood, and they were cast upon the earth: and the third part of trees was burnt up, and all green grass was burnt up.

8 And the second angel sounded, and as it were a great mountain burning with fire was cast into the sea: and the third part of the sea became blood;

9 And the third part of the creatures which were in the sea, and had life, died; and the third part of the ships were destroyed.

10 And the third angel sounded, and there fell a great star from heaven, burning as it were a lamp, and it fell upon the third part of the rivers, and upon the fountains of waters;

¹¹ And the name of the star is called Wormwood: and the third part of the waters became wormwood; and many men died of the waters, because they were made bitter.

¹² And the fourth angel sounded, and the third part of the sun was smitten, and the third part of the moon, and the third part of the stars; so as the third part of them was darkened, and the day shone not for a third part of it, and the night likewise.

¹³ And I beheld, and heard an angel flying through the midst of heaven, saying with a loud voice, Woe, woe, woe, to the inhabiters of the earth by reason of the other voices of the trumpet of the three angels, which are yet to sound!

Revelation Chapter 9:1-21

And the fifth angel sounded, and I saw a star fall from heaven unto the earth: and to him was given the key of the bottomless pit.

² And he opened the bottomless pit; and there arose a smoke out of the pit, as the smoke of a great furnace; and the sun and the air were darkened by reason of the smoke of the pit.

³ And there came out of the smoke locusts upon the earth: and unto them was given power, as the scorpions of the earth have power.

4 And it was commanded them that they should not hurt the grass of the earth, neither any green thing, neither any tree; but only those men which have not the seal of God in their foreheads.

5 And to them it was given that they should not kill them, but that they should be tormented five months: and their torment was as the torment of a scorpion, when he striketh a man.

6 And in those days shall men seek death, and shall not find it; and shall desire to die, and death shall flee from them.

7 And the shapes of the locusts were like unto horses prepared unto battle; and on their heads were as it were crowns like gold, and their faces were as the faces of men.

8 And they had hair as the hair of women, and their teeth were as the teeth of lions.

9 And they had breastplates, as it were breastplates of iron; and the sound of their wings was as the sound of chariots of many horses running to battle.

10 And they had tails like unto scorpions, and there were stings in their tails: and their power was to hurt men five months.

11 And they had a king over them, which is the angel of the bottomless pit, whose name in the Hebrew tongue is Abaddon, but in the Greek tongue hath his name Apollyon.

¹² One woe is past; and, behold, there come two woes more hereafter.

¹³ And the sixth angel sounded, and I heard a voice from the four horns of the golden altar which is before God,

¹⁴ Saying to the sixth angel which had the trumpet, Loose the four angels which are bound in the great river Euphrates.

¹⁵ And the four angels were loosed, which were prepared for an hour, and a day, and a month, and a year, for to slay the third part of men.

¹⁶ And the number of the army of the horsemen were two hundred thousand thousand: and I heard the number of them.

¹⁷ And thus I saw the horses in the vision, and them that sat on them, having breastplates of fire, and of jacinth, and brimstone: and the heads of the horses were as the heads of lions; and out of their mouths issued fire and smoke and brimstone.

¹⁸ By these three was the third part of men killed, by the fire, and by the smoke, and by the brimstone, which issued out of their mouths.

¹⁹ For their power is in their mouth, and in their tails: for their tails were like unto serpents, and had heads, and with them they do hurt.

20 And the rest of the men which were not killed by these plagues yet repented not of the works of their hands, that they should not worship devils, and idols of gold, and silver, and brass, and stone, and of wood: which neither can see, nor hear, nor walk:

21 Neither repented they of their murders, nor of their sorceries, nor of their fornication, nor of their thefts.

Revelation Chapter 10:1-11

And I saw another mighty angel come down from heaven, clothed with a cloud: and a rainbow was upon his head, and his face was as it were the sun, and his feet as pillars of fire:

2 And he had in his hand a little book open: and he set his right foot upon the sea, and his left foot on the earth,

3 And cried with a loud voice, as when a lion roareth: and when he had cried, seven thunders uttered their voices.

4 And when the seven thunders had uttered their voices, I was about to write: and I heard a voice from heaven saying unto me, Seal up those things which the seven thunders uttered, and write them not.

5 And the angel which I saw stand upon the sea and upon the earth lifted up his hand to heaven,

⁶ And sware by him that liveth for ever and ever, who created heaven, and the things that therein are, and the earth, and the things that therein are, and the sea, and the things which are therein, that there should be time no longer:

⁷ But in the days of the voice of the seventh angel, when he shall begin to sound, the mystery of God should be finished, as he hath declared to his servants the prophets.

⁸ And the voice which I heard from heaven spake unto me again, and said, Go and take the little book which is open in the hand of the angel which standeth upon the sea and upon the earth.

⁹ And I went unto the angel, and said unto him, Give me the little book. And he said unto me, Take it, and eat it up; and it shall make thy belly bitter, but it shall be in thy mouth sweet as honey.

¹⁰ And I took the little book out of the angel's hand, and ate it up; and it was in my mouth sweet as honey: and as soon as I had eaten it, my belly was bitter.

¹¹ And he said unto me, Thou must prophesy again before many peoples, and nations, and tongues, and kings.

Revelation Chapter 11:1-19

And there was given me a reed like unto a rod: and the angel stood, saying, Rise, and measure the temple of God, and the altar, and them that worship therein.

Visitors From Another Dimension

2 But the court which is without the temple leave out, and measure it not; for it is given unto the Gentiles: and the holy city shall they tread under foot forty and two months.

3 And I will give power unto my two witnesses, and they shall prophesy a thousand two hundred and threescore days, clothed in sackcloth.

4 These are the two olive trees, and the two candlesticks standing before the God of the earth.

5 And if any man will hurt them, fire proceedeth out of their mouth, and devoureth their enemies: and if any man will hurt them, he must in this manner be killed.

6 These have power to shut heaven, that it rain not in the days of their prophecy: and have power over waters to turn them to blood, and to smite the earth with all plagues, as often as they will.

7 And when they shall have finished their testimony, the beast that ascendeth out of the bottomless pit shall make war against them, and shall overcome them, and kill them.

8 And their dead bodies shall lie in the street of the great city, which spiritually is called Sodom and Egypt, where also our Lord was crucified.

9 And they of the people and kindreds and tongues and nations shall see their dead bodies three days and an half, and shall not suffer their dead bodies to be put in graves.

¹⁰ And they that dwell upon the earth shall rejoice over them, and make merry, and shall send gifts one to another; because these two prophets tormented them that dwelt on the earth.

¹¹ And after three days and an half the spirit of life from God entered into them, and they stood upon their feet; and great fear fell upon them which saw them.

¹² And they heard a great voice from heaven saying unto them, Come up hither. And they ascended up to heaven in a cloud; and their enemies beheld them.

¹³ And the same hour was there a great earthquake, and the tenth part of the city fell, and in the earthquake were slain of men seven thousand: and the remnant were affrighted, and gave glory to the God of heaven.

¹⁴ The second woe is past; and, behold, the third woe cometh quickly.

¹⁵ And the seventh angel sounded; and there were great voices in heaven, saying, The kingdoms of this world are become the kingdoms of our Lord, and of his Christ; and he shall reign for ever and ever.

¹⁶ And the four and twenty elders, which sat before God on their seats, fell upon their faces, and worshipped God,

¹⁷ Saying, We give thee thanks, O Lord God Almighty, which art, and wast, and art to come; because thou hast taken to thee thy great power, and hast reigned.

18 And the nations were angry, and thy wrath is come, and the time of the dead, that they should be judged, and that thou shouldest give reward unto thy servants the prophets, and to the saints, and them that fear thy name, small and great; and shouldest destroy them which destroy the earth.

19 And the temple of God was opened in heaven, and there was seen in his temple the ark of his testament: and there were lightnings, and voices, and thunderings, and an earthquake, and great hail

Revelation Chapter 12:1-17

And there appeared a great wonder in heaven; a woman clothed with the sun, and the moon under her feet, and upon her head a crown of twelve stars:

2 And she being with child cried, travailing in birth, and pained to be delivered.

3 And there appeared another wonder in heaven; and behold a great red dragon, having seven heads and ten horns, and seven crowns upon his heads.

4 And his tail drew the third part of the stars of heaven, and did cast them to the earth: and the dragon stood before the woman which was ready to be delivered, for to devour her child as soon as it was born.

5 And she brought forth a man child, who was to rule all nations with a rod of iron: and her child was caught up unto God, and to his throne.

⁶ And the woman fled into the wilderness, where she hath a place prepared of God, that they should feed her there a thousand two hundred and threescore days.

⁷ And there was war in heaven: Michael and his angels fought against the dragon; and the dragon fought and his angels,

⁸ And prevailed not; neither was their place found any more in heaven.

⁹ And the great dragon was cast out, that old serpent, called the Devil, and Satan, which deceiveth the whole world: he was cast out into the earth, and his angels were cast out with him.

¹⁰ And I heard a loud voice saying in heaven, Now is come salvation, and strength, and the kingdom of our God, and the power of his Christ: for the accuser of our brethren is cast down, which accused them before our God day and night.

¹¹ And they overcame him by the blood of the Lamb, and by the word of their testimony; and they loved not their lives unto the death.

¹² Therefore rejoice, ye heavens, and ye that dwell in them. Woe to the inhabiters of the earth and of the sea! for the devil is come down unto you, having great wrath, because he knoweth that he hath but a short time.

¹³ And when the dragon saw that he was cast unto the earth, he persecuted the woman which brought forth the man child.

14 And to the woman were given two wings of a great eagle, that she might fly into the wilderness, into her place, where she is nourished for a time, and times, and half a time, from the face of the serpent.

15 And the serpent cast out of his mouth water as a flood after the woman, that he might cause her to be carried away of the flood.

16 And the earth helped the woman, and the earth opened her mouth, and swallowed up the flood which the dragon cast out of his mouth.

17 And the dragon was wroth with the woman, and went to make war with the remnant of her seed, which keep the commandments of God, and have the testimony of Jesus Christ.

Revelation Chapter 13:1-18

And I stood upon the sand of the sea, and saw a beast rise up out of the sea, having seven heads and ten horns, and upon his horns ten crowns, and upon his heads the name of blasphemy.

2 And the beast which I saw was like unto a leopard, and his feet were as the feet of a bear, and his mouth as the mouth of a lion: and the dragon gave him his power, and his seat, and great authority.

3 And I saw one of his heads as it were wounded to death; and his deadly wound was healed: and all the world wondered after the beast.

⁴ And they worshipped the dragon which gave power unto the beast: and they worshipped the beast, saying, Who is like unto the beast? who is able to make war with him?

⁵ And there was given unto him a mouth speaking great things and blasphemies; and power was given unto him to continue forty and two months.

⁶ And he opened his mouth in blasphemy against God, to blaspheme his name, and his tabernacle, and them that dwell in heaven.

⁷ And it was given unto him to make war with the saints, and to overcome them: and power was given him over all kindreds, and tongues, and nations.

⁸ And all that dwell upon the earth shall worship him, whose names are not written in the book of life of the Lamb slain from the foundation of the world.

⁹ If any man have an ear, let him hear.

¹⁰ He that leadeth into captivity shall go into captivity: he that killeth with the sword must be killed with the sword. Here is the patience and the faith of the saints.

¹¹ And I beheld another beast coming up out of the earth; and he had two horns like a lamb, and he spake as a dragon.

¹² And he exerciseth all the power of the first beast before him, and causeth the earth and them which dwell therein to worship the first beast, whose deadly wound was healed.

¹³ And he doeth great wonders, so that he maketh fire come down from heaven on the earth in the sight of men,

¹⁴ And deceiveth them that dwell on the earth by the means of those miracles which he had power to do in the sight of the beast; saying to them that dwell on the earth, that they should make an image to the beast, which had the wound by a sword, and did live.

¹⁵ And he had power to give life unto the image of the beast, that the image of the beast should both speak, and cause that as many as would not worship the image of the beast should be killed.

¹⁶ And he causeth all, both small and great, rich and poor, free and bond, to receive a mark in their right hand, or in their foreheads:

¹⁷ And that no man might buy or sell, save he that had the mark, or the name of the beast, or the number of his name.

¹⁸ Here is wisdom. Let him that hath understanding count the number of the beast: for it is the number of a man; and his number is Six hundred threescore and six.

Revelation Chapter 14:1-20

And I looked, and, lo, a Lamb stood on the mount Sion, and with him an hundred forty and four thousand, having his Father's name written in their foreheads.

² And I heard a voice from heaven, as the voice of many waters, and as the voice of a great thunder: and I heard the voice of harpers harping with their harps:

³ And they sung as it were a new song before the throne, and before the four beasts, and the elders: and no man could learn that song but the hundred and forty and four thousand, which were redeemed from the earth.

⁴ These are they which were not defiled with women; for they are virgins. These are they which follow the Lamb whithersoever he goeth. These were redeemed from among men, being the firstfruits unto God and to the Lamb.

⁵ And in their mouth was found no guile: for they are without fault before the throne of God.

⁶ And I saw another angel fly in the midst of heaven, having the everlasting gospel to preach unto them that dwell on the earth, and to every nation, and kindred, and tongue, and people,

⁷ Saying with a loud voice, Fear God, and give glory to him; for the hour of his judgment is come: and worship him that made heaven, and earth, and the sea, and the fountains of waters.

⁸ And there followed another angel, saying, Babylon is fallen, is fallen, that great city, because she made all nations drink of the wine of the wrath of her fornication.

⁹ And the third angel followed them, saying with a loud voice, If any man worship the beast and his image, and receive his mark in his forehead, or in his hand,

¹⁰ The same shall drink of the wine of the wrath of God, which is poured out without mixture into the cup of his indignation; and he shall be tormented with fire and brimstone in the presence of the holy angels, and in the presence of the Lamb:

¹¹ And the smoke of their torment ascendeth up for ever and ever: and they have no rest day nor night, who worship the beast and his image, and whosoever receiveth the mark of his name.

¹² Here is the patience of the saints: here are they that keep the commandments of God, and the faith of Jesus.

¹³ And I heard a voice from heaven saying unto me, Write, Blessed are the dead which die in the Lord from henceforth: Yea, saith the Spirit, that they may rest from their labours; and their works do follow them.

¹⁴ And I looked, and behold a white cloud, and upon the cloud one sat like unto the Son of man, having on his head a golden crown, and in his hand a sharp sickle.

¹⁵ And another angel came out of the temple, crying with a loud voice to him that sat on the cloud, Thrust in thy sickle, and reap: for the time is come for thee to reap; for the harvest of the earth is ripe.

¹⁶ And he that sat on the cloud thrust in his sickle on the earth; and the earth was reaped.

¹⁷ And another angel came out of the temple which is in heaven, he also having a sharp sickle.

¹⁸ And another angel came out from the altar, which had power over fire; and cried with a loud cry to him that had the sharp sickle, saying, Thrust in thy sharp sickle, and gather the clusters of the vine of the earth; for her grapes are fully ripe.

¹⁹ And the angel thrust in his sickle into the earth, and gathered the vine of the earth, and cast it into the great winepress of the wrath of God.

²⁰ And the winepress was trodden without the city, and blood came out of the winepress, even unto the horse bridles, by the space of a thousand and six hundred furlongs.

Revelation Chapter 15:1-8

And I saw another sign in heaven, great and marvellous, seven angels having the seven last plagues; for in them is filled up the wrath of God.

² And I saw as it were a sea of glass mingled with fire: and them that had gotten the victory over the beast, and over his image, and over his mark, and over the number of his name, stand on the sea of glass, having the harps of God.

³ And they sing the song of Moses the servant of God, and the song of the Lamb, saying, Great and marvellous are thy works, Lord God Almighty; just and true are thy ways, thou King of saints.

⁴ Who shall not fear thee, O Lord, and glorify thy name? for thou only art holy: for all nations shall come and worship before thee; for thy judgments are made manifest.

⁵ And after that I looked, and, behold, the temple of the tabernacle of the testimony in heaven was opened:

⁶ And the seven angels came out of the temple, having the seven plagues, clothed in pure and white linen, and having their breasts girded with golden girdles.

⁷ And one of the four beasts gave unto the seven angels seven golden vials full of the wrath of God, who liveth for ever and ever.

⁸ And the temple was filled with smoke from the glory of God, and from his power; and no man was able to enter into the temple, till the seven plagues of the seven angels were fulfilled.

Revelation Chapter 16:1-21

And I heard a great voice out of the temple saying to the seven angels, Go your ways, and pour out the vials of the wrath of God upon the earth.

Visitors From Another Dimension

2 And the first went, and poured out his vial upon the earth; and there fell a noisome and grievous sore upon the men which had the mark of the beast, and upon them which worshipped his image.

3 And the second angel poured out his vial upon the sea; and it became as the blood of a dead man: and every living soul died in the sea.

4 And the third angel poured out his vial upon the rivers and fountains of waters; and they became blood.

5 And I heard the angel of the waters say, Thou art righteous, O Lord, which art, and wast, and shalt be, because thou hast judged thus.

6 For they have shed the blood of saints and prophets, and thou hast given them blood to drink; for they are worthy.

7 And I heard another out of the altar say, Even so, Lord God Almighty, true and righteous are thy judgments.

8 And the fourth angel poured out his vial upon the sun; and power was given unto him to scorch men with fire.

9 And men were scorched with great heat, and blasphemed the name of God, which hath power over these plagues: and they repented not to give him glory.

10 And the fifth angel poured out his vial upon the seat of the beast; and his kingdom was full of darkness; and they gnawed their tongues for pain,

¹¹ And blasphemed the God of heaven because of their pains and their sores, and repented not of their deeds.

¹² And the sixth angel poured out his vial upon the great river Euphrates; and the water thereof was dried up, that the way of the kings of the east might be prepared.

¹³ And I saw three unclean spirits like frogs come out of the mouth of the dragon, and out of the mouth of the beast, and out of the mouth of the false prophet.

¹⁴ For they are the spirits of devils, working miracles, which go forth unto the kings of the earth and of the whole world, to gather them to the battle of that great day of God Almighty.

¹⁵ Behold, I come as a thief. Blessed is he that watcheth, and keepeth his garments, lest he walk naked, and they see his shame.

¹⁶ And he gathered them together into a place called in the Hebrew tongue Armageddon.

¹⁷ And the seventh angel poured out his vial into the air; and there came a great voice out of the temple of heaven, from the throne, saying, It is done.

¹⁸ And there were voices, and thunders, and lightnings; and there was a great earthquake, such as was not since men were upon the earth, so mighty an earthquake, and so great.

¹⁹ And the great city was divided into three parts, and the cities of the nations fell: and great Babylon came in remembrance before God, to give unto her the cup of the wine of the fierceness of his wrath.

²⁰ And every island fled away, and the mountains were not found.

²¹ And there fell upon men a great hail out of heaven, every stone about the weight of a talent: and men blasphemed God because of the plague of the hail; for the plague thereof was exceeding great.

Revelation Chapter 17:1-18

And there came one of the seven angels which had the seven vials, and talked with me, saying unto me, Come hither; I will shew unto thee the judgment of the great whore that sitteth upon many waters:

² With whom the kings of the earth have committed fornication, and the inhabitants of the earth have been made drunk with the wine of her fornication.

³ So he carried me away in the spirit into the wilderness: and I saw a woman sit upon a scarlet coloured beast, full of names of blasphemy, having seven heads and ten horns.

⁴ And the woman was arrayed in purple and scarlet colour, and decked with gold and precious stones and pearls, having a golden cup in her hand full of abominations and filthiness of her fornication:

5 And upon her forehead was a name written, Mystery, Babylon The Great, The Mother Of Harlots And Abominations Of The Earth.

6 And I saw the woman drunken with the blood of the saints, and with the blood of the martyrs of Jesus: and when I saw her, I wondered with great admiration.

7 And the angel said unto me, Wherefore didst thou marvel? I will tell thee the mystery of the woman, and of the beast that carrieth her, which hath the seven heads and ten horns.

8 The beast that thou sawest was, and is not; and shall ascend out of the bottomless pit, and go into perdition: and they that dwell on the earth shall wonder, whose names were not written in the book of life from the foundation of the world, when they behold the beast that was, and is not, and yet is.

9 And here is the mind which hath wisdom. The seven heads are seven mountains, on which the woman sitteth.

10 And there are seven kings: five are fallen, and one is, and the other is not yet come; and when he cometh, he must continue a short space.

11 And the beast that was, and is not, even he is the eighth, and is of the seven, and goeth into perdition.

12 And the ten horns which thou sawest are ten kings, which have received no kingdom as yet; but receive power as kings one hour with the beast.

¹³ These have one mind, and shall give their power and strength unto the beast.

¹⁴ These shall make war with the Lamb, and the Lamb shall overcome them: for he is Lord of lords, and King of kings: and they that are with him are called, and chosen, and faithful.

¹⁵ And he saith unto me, The waters which thou sawest, where the whore sitteth, are peoples, and multitudes, and nations, and tongues.

¹⁶ And the ten horns which thou sawest upon the beast, these shall hate the whore, and shall make her desolate and naked, and shall eat her flesh, and burn her with fire.

¹⁷ For God hath put in their hearts to fulfil his will, and to agree, and give their kingdom unto the beast, until the words of God shall be fulfilled.

¹⁸ And the woman which thou sawest is that great city, which reigneth over the kings of the earth.

Revelation Chapter 18:1-24

And after these things I saw another angel come down from heaven, having great power; and the earth was lightened with his glory.

² And he cried mightily with a strong voice, saying, Babylon the great is fallen, is fallen, and is become the habitation of devils, and the hold of every foul spirit, and a cage of every unclean and hateful bird.

3 For all nations have drunk of the wine of the wrath of her fornication, and the kings of the earth have committed fornication with her, and the merchants of the earth are waxed rich through the abundance of her delicacies.

4 And I heard another voice from heaven, saying, Come out of her, my people, that ye be not partakers of her sins, and that ye receive not of her plagues.

5 For her sins have reached unto heaven, and God hath remembered her iniquities.

6 Reward her even as she rewarded you, and double unto her double according to her works: in the cup which she hath filled fill to her double.

7 How much she hath glorified herself, and lived deliciously, so much torment and sorrow give her: for she saith in her heart, I sit a queen, and am no widow, and shall see no sorrow.

8 Therefore shall her plagues come in one day, death, and mourning, and famine; and she shall be utterly burned with fire: for strong is the Lord God who judgeth her.

9 And the kings of the earth, who have committed fornication and lived deliciously with her, shall bewail her, and lament for her, when they shall see the smoke of her burning,

10 Standing afar off for the fear of her torment, saying, Alas, alas that great city Babylon, that mighty city! for in one hour is thy judgment come.

¹¹ And the merchants of the earth shall weep and mourn over her; for no man buyeth their merchandise any more:

¹² The merchandise of gold, and silver, and precious stones, and of pearls, and fine linen, and purple, and silk, and scarlet, and all thyine wood, and all manner vessels of ivory, and all manner vessels of most precious wood, and of brass, and iron, and marble,

¹³ And cinnamon, and odours, and ointments, and frankincense, and wine, and oil, and fine flour, and wheat, and beasts, and sheep, and horses, and chariots, and slaves, and souls of men.

¹⁴ And the fruits that thy soul lusted after are departed from thee, and all things which were dainty and goodly are departed from thee, and thou shalt find them no more at all.

¹⁵ The merchants of these things, which were made rich by her, shall stand afar off for the fear of her torment, weeping and wailing,

¹⁶ And saying, Alas, alas that great city, that was clothed in fine linen, and purple, and scarlet, and decked with gold, and precious stones, and pearls!

¹⁷ For in one hour so great riches is come to nought. And every shipmaster, and all the company in ships, and sailors, and as many as trade by sea, stood afar off,

¹⁸ And cried when they saw the smoke of her burning, saying, What city is like unto this great city!

¹⁹ And they cast dust on their heads, and cried, weeping and wailing, saying, Alas, alas that great city, wherein were made rich all that had ships in the sea by reason of her costliness! for in one hour is she made desolate.

²⁰ Rejoice over her, thou heaven, and ye holy apostles and prophets; for God hath avenged you on her.

²¹ And a mighty angel took up a stone like a great millstone, and cast it into the sea, saying, Thus with violence shall that great city Babylon be thrown down, and shall be found no more at all.

²² And the voice of harpers, and musicians, and of pipers, and trumpeters, shall be heard no more at all in thee; and no craftsman, of whatsoever craft he be, shall be found any more in thee; and the sound of a millstone shall be heard no more at all in thee;

²³ And the light of a candle shall shine no more at all in thee; and the voice of the bridegroom and of the bride shall be heard no more at all in thee: for thy merchants were the great men of the earth; for by thy sorceries were all nations deceived.

²⁴ And in her was found the blood of prophets, and of saints, and of all that were slain upon the earth.

Revelation Chapter 19:1-21

And after these things I heard a great voice of much people in heaven, saying, Alleluia; Salvation, and glory, and honour, and power, unto the Lord our God:

2 For true and righteous are his judgments: for he hath judged the great whore, which did corrupt the earth with her fornication, and hath avenged the blood of his servants at her hand.

3 And again they said, Alleluia And her smoke rose up for ever and ever.

4 And the four and twenty elders and the four beasts fell down and worshipped God that sat on the throne, saying, Amen; Alleluia.

5 And a voice came out of the throne, saying, Praise our God, all ye his servants, and ye that fear him, both small and great.

6 And I heard as it were the voice of a great multitude, and as the voice of many waters, and as the voice of mighty thunderings, saying, Alleluia: for the Lord God omnipotent reigneth.

7 Let us be glad and rejoice, and give honour to him: for the marriage of the Lamb is come, and his wife hath made herself ready.

8 And to her was granted that she should be arrayed in fine linen, clean and white: for the fine linen is the righteousness of saints.

9 And he saith unto me, Write, Blessed are they which are called unto the marriage supper of the Lamb. And he saith unto me, These are the true sayings of God.

Visitors From Another Dimension

¹⁰ And I fell at his feet to worship him. And he said unto me, See thou do it not: I am thy fellowservant, and of thy brethren that have the testimony of Jesus: worship God: for the testimony of Jesus is the spirit of prophecy.

¹¹ And I saw heaven opened, and behold a white horse; and he that sat upon him was called Faithful and True, and in righteousness he doth judge and make war.

¹² His eyes were as a flame of fire, and on his head were many crowns; and he had a name written, that no man knew, but he himself.

¹³ And he was clothed with a vesture dipped in blood: and his name is called The Word of God.

¹⁴ And the armies which were in heaven followed him upon white horses, clothed in fine linen, white and clean.

¹⁵ And out of his mouth goeth a sharp sword, that with it he should smite the nations: and he shall rule them with a rod of iron: and he treadeth the winepress of the fierceness and wrath of Almighty God.

¹⁶ And he hath on his vesture and on his thigh a name written, King Of Kings, And Lord Of Lords.

¹⁷ And I saw an angel standing in the sun; and he cried with a loud voice, saying to all the fowls that fly in the midst of heaven, Come and gather yourselves together unto the supper of the great God;

¹⁸ That ye may eat the flesh of kings, and the flesh of captains, and the flesh of mighty men, and the flesh of horses, and of them that sit on them, and the flesh of all men, both free and bond, both small and great.

¹⁹ And I saw the beast, and the kings of the earth, and their armies, gathered together to make war against him that sat on the horse, and against his army.

²⁰ And the beast was taken, and with him the false prophet that wrought miracles before him, with which he deceived them that had received the mark of the beast, and them that worshipped his image. These both were cast alive into a lake of fire burning with brimstone.

²¹ And the remnant were slain with the sword of him that sat upon the horse, which sword proceeded out of his mouth: and all the fowls were filled with their flesh.

Revelation Chapter 20:1-15

And I saw an angel come down from heaven, having the key of the bottomless pit and a great chain in his hand.

² And he laid hold on the dragon, that old serpent, which is the Devil, and Satan, and bound him a thousand years,

³ And cast him into the bottomless pit, and shut him up, and set a seal upon him, that he should deceive the nations no more, till the thousand years should be fulfilled: and after that he must be loosed a little season.

Visitors From Another Dimension

⁴ And I saw thrones, and they sat upon them, and judgment was given unto them: and I saw the souls of them that were beheaded for the witness of Jesus, and for the word of God, and which had not worshipped the beast, neither his image, neither had received his mark upon their foreheads, or in their hands; and they lived and reigned with Christ a thousand years.

⁵ But the rest of the dead lived not again until the thousand years were finished. This is the first resurrection.

⁶ Blessed and holy is he that hath part in the first resurrection: on such the second death hath no power, but they shall be priests of God and of Christ, and shall reign with him a thousand years.

⁷ And when the thousand years are expired, Satan shall be loosed out of his prison,

⁸ And shall go out to deceive the nations which are in the four quarters of the earth, Gog, and Magog, to gather them together to battle: the number of whom is as the sand of the sea.

⁹ And they went up on the breadth of the earth, and compassed the camp of the saints about, and the beloved city: and fire came down from God out of heaven, and devoured them.

¹⁰ And the devil that deceived them was cast into the lake of fire and brimstone, where the beast and the false prophet are, and shall be tormented day and night for ever and ever.

¹¹ And I saw a great white throne, and him that sat on it, from whose face the earth and the heaven fled away; and there was found no place for them.

¹² And I saw the dead, small and great, stand before God; and the books were opened: and another book was opened, which is the book of life: and the dead were judged out of those things which were written in the books, according to their works.

¹³ And the sea gave up the dead which were in it; and death and hell delivered up the dead which were in them: and they were judged every man according to their works.

¹⁴ And death and hell were cast into the lake of fire. This is the second death.

¹⁵ And whosoever was not found written in the book of life was cast into the lake of fire.

Revelation Chapter 21:1-27

And I saw a new heaven and a new earth: for the first heaven and the first earth were passed away; and there was no more sea.

² And I John saw the holy city, new Jerusalem, coming down from God out of heaven, prepared as a bride adorned for her husband.

³ And I heard a great voice out of heaven saying, Behold, the tabernacle of God is with men, and he will dwell with them, and they shall be his people, and God himself shall be with them, and be their God.

⁴ And God shall wipe away all tears from their eyes; and there shall be no more death, neither sorrow, nor crying, neither shall there be any more pain: for the former things are passed away.

⁵ And he that sat upon the throne said, Behold, I make all things new. And he said unto me, Write: for these words are true and faithful.

⁶ And he said unto me, It is done. I am Alpha and Omega, the beginning and the end. I will give unto him that is athirst of the fountain of the water of life freely.

⁷ He that overcometh shall inherit all things; and I will be his God, and he shall be my son.

⁸ But the fearful, and unbelieving, and the abominable, and murderers, and whoremongers, and sorcerers, and idolaters, and all liars, shall have their part in the lake which burneth with fire and brimstone: which is the second death.

⁹ And there came unto me one of the seven angels which had the seven vials full of the seven last plagues, and talked with me, saying, Come hither, I will shew thee the bride, the Lamb's wife.

¹⁰ And he carried me away in the spirit to a great and high mountain, and shewed me that great city, the holy Jerusalem, descending out of heaven from God,

¹¹ Having the glory of God: and her light was like unto a stone most precious, even like a jasper stone, clear as crystal;

¹² And had a wall great and high, and had twelve gates, and at the gates twelve angels, and names written thereon, which are the names of the twelve tribes of the children of Israel:

¹³ On the east three gates; on the north three gates; on the south three gates; and on the west three gates.

¹⁴ And the wall of the city had twelve foundations, and in them the names of the twelve apostles of the Lamb.

¹⁵ And he that talked with me had a golden reed to measure the city, and the gates thereof, and the wall thereof.

¹⁶ And the city lieth foursquare, and the length is as large as the breadth: and he measured the city with the reed, twelve thousand furlongs. The length and the breadth and the height of it are equal.

¹⁷ And he measured the wall thereof, an hundred and forty and four cubits, according to the measure of a man, that is, of the angel.

¹⁸ And the building of the wall of it was of jasper: and the city was pure gold, like unto clear glass.

¹⁹ And the foundations of the wall of the city were garnished with all manner of precious stones. The first foundation was jasper; the second, sapphire; the third, a chalcedony; the fourth, an emerald;

²⁰ The fifth, sardonyx; the sixth, sardius; the seventh, chrysolyte; the eighth, beryl; the ninth, a topaz; the tenth, a chrysoprasus; the eleventh, a jacinth; the twelfth, an amethyst.

²¹ And the twelve gates were twelve pearls: every several gate was of one pearl: and the street of the city was pure gold, as it were transparent glass.

²² And I saw no temple therein: for the Lord God Almighty and the Lamb are the temple of it.

²³ And the city had no need of the sun, neither of the moon, to shine in it: for the glory of God did lighten it, and the Lamb is the light thereof.

²⁴ And the nations of them which are saved shall walk in the light of it: and the kings of the earth do bring their glory and honour into it.

²⁵ And the gates of it shall not be shut at all by day: for there shall be no night there.

²⁶ And they shall bring the glory and honour of the nations into it.

²⁷ And there shall in no wise enter into it any thing that defileth, neither whatsoever worketh abomination, or maketh a lie: but they which are written in the Lamb's book of life.

Revelation Chapter 22:1-21

And he shewed me a pure river of water of life, clear as crystal, proceeding out of the throne of God and of the Lamb.

Visitors From Another Dimension

² In the midst of the street of it, and on either side of the river, was there the tree of life, which bare twelve manner of fruits, and yielded her fruit every month: and the leaves of the tree were for the healing of the nations.

³ And there shall be no more curse: but the throne of God and of the Lamb shall be in it; and his servants shall serve him:

⁴ And they shall see his face; and his name shall be in their foreheads.

⁵ And there shall be no night there; and they need no candle, neither light of the sun; for the Lord God giveth them light: and they shall reign for ever and ever.

⁶ And he said unto me, These sayings are faithful and true: and the Lord God of the holy prophets sent his angel to shew unto his servants the things which must shortly be done.

⁷ Behold, I come quickly: blessed is he that keepeth the sayings of the prophecy of this book.

⁸ And I John saw these things, and heard them. And when I had heard and seen, I fell down to worship before the feet of the angel which shewed me these things.

⁹ Then saith he unto me, See thou do it not: for I am thy fellowservant, and of thy brethren the prophets, and of them which keep the sayings of this book: worship God.

Visitors From Another Dimension

¹⁰ And he saith unto me, Seal not the sayings of the prophecy of this book: for the time is at hand.

¹¹ He that is unjust, let him be unjust still: and he which is filthy, let him be filthy still: and he that is righteous, let him be righteous still: and he that is holy, let him be holy still.

¹² And, behold, I come quickly; and my reward is with me, to give every man according as his work shall be.

¹³ I am Alpha and Omega, the beginning and the end, the first and the last.

¹⁴ Blessed are they that do his commandments, that they may have right to the tree of life, and may enter in through the gates into the city.

¹⁵ For without are dogs, and sorcerers, and whoremongers, and murderers, and idolaters, and whosoever loveth and maketh a lie.

¹⁶ I Jesus have sent mine angel to testify unto you these things in the churches. I am the root and the offspring of David, and the bright and morning star.

¹⁷ And the Spirit and the bride say, Come. And let him that heareth say, Come. And let him that is athirst come. And whosoever will, let him take the water of life freely.

¹⁸ For I testify unto every man that heareth the words of the prophecy of this book, If any man shall add unto these things, God shall add unto him the plagues that are written in this book:

[19] And if any man shall take away from the words of the book of this prophecy, God shall take away his part out of the book of life, and out of the holy city, and from the things which are written in this book.

[20] He which testifieth these things saith, Surely I come quickly. Amen. Even so, come, Lord Jesus.

[21] The grace of our Lord Jesus Christ be with you all. Amen.